T0189337

Industrial IoT

Ismail Butun
Editor

Industrial IoT

Challenges, Design Principles, Applications, and Security

Springer

Editor
Ismail Butun
Chalmers University of Technology
Göteborg, Sweden

ISBN 978-3-030-42502-9 ISBN 978-3-030-42500-5 (eBook)
https://doi.org/10.1007/978-3-030-42500-5

This Springer imprint is published by the registered company Springer Nature Switzerland AG
The registered company address is: Gewerbestrasse 11, 6330 Cham, Switzerland

In memory of my grandparents
and
my father Orhan Bütün . . .

To my mother Emine Bütün

İsmail Bütün

Foreword

I am honored to write this foreword for the book: "Industrial IoT: Challenges, Design Principles, Applications, and Security." The Editor of this book, Dr. Ismail Butun, is a well-known researcher with impactful contributions in wireless communications, computer networks, and network security for the past several years. He published more than 35 scientific articles which were noticed by the research community. His publications have already received more than 1000 citations along with an H-index of 12. He demonstrates his knowledge in this edited book. I find this book very useful for academicians and practitioners in the industry.

As the industrial revolution (a.k.a. Industry 4.0) continues at full pace, it is indispensable to include all the benefits offered by the wireless technology, as it is also evolving at a very fast pace. Besides, as the autonomous robots are invading the factory floors at a rapid rate, they constitute another fleet of things in the IoT to be wirelessly connected to each other and to the control center. Moreover, after several decades of design and development, the smart grid and micro-grid technologies have evolved from the traditional electric grid to the point where they include remote monitoring and control, along with smart meters, sending gigabytes of information per hour to the control center. Further, using the 4G/5G cellular and low-power WAN wireless technologies, the Internet of Things (IoT) such as NB-IoT and LoRaWAN are more flexible, deployable, scalable, and reachable than ever before.

With all these advances, the Industrial IoT (so-called IIoT) concept is evolving and comprises the primary focus of this book. It introduces all the recent technologies devised for the industrial networks, IoT, and IIoT domains, from factory floor deployments to in-house applications. Therefore, it is not only for experts and academicians in the field but also for the beginners and practitioners in the industry.

This book covers a wide range of topics, including but not limited to, the digital twin, IoT-based industrial indoor/outdoor lighting systems, wireless communications challenges and opportunities, decentralized computing, fog/edge/cloud computing, data streaming, cyber-security, and intrusion detection.

Georgia Tech
Atlanta, GA, USA
January 2020

Prof. Ian F. Akyildiz

Preface

We are all living in a connected world and Cisco predicts 500 Billion things of the IoT to be further included in this connection; meaning more automation, remote access and control to be infused in our everyday routines.

This book, "Industrial IoT: Challenges, Design Principles, Applications, and Security," aims at presenting the recent developments in the fields of industrial networks, IoT, and IIoT domains. For this book, 18 chapter proposals were submitted from the academicians and practitioners in the field. After careful reviews, six chapters were finally accepted to be included in this book with an acceptance rate of 33%.

Since readers of this book are likely to come from various backgrounds, being aware of the implicit structure of this book might be helpful. The contributed chapters in this book cover a broad range of topics related to IIoT networks, including wired/wireless communication technologies, industrial applications, cyber-security, and intrusion detection. The book consists of three parts and six chapters, which I find a convenient way of presenting the overall material:

Part I consists of Chaps. 1 and 2 to introduce preliminaries, design principles, and challenges of the IIoT. Chapter 1 introduces an overview of most of the networking, communication, and ICT technologies available in the industrial networks, whereas Chap. 2 presents wireless communication technologies that apply to IIoT and also discussed their unique challenges.

Part II consists of Chaps. 3 and 4 to introduce automation trends and applications of IIoT. Chapter 3 is dedicated to IoT-driven advances in industrial and commercial smart buildings, especially new advances at IoT-based industrial indoor/outdoor lighting systems. Chapter 4 introduces the automation trends in industrial networks and IIoT, including the most famous digital twin concept.

Part III consists of Chaps. 5 and 6 to stress on the cyber-security of IIoT. Chapter 5 presents the security of IIoT networks, especially within the decentralized cloud computing settings of the IIoT. Finally, Chap. 6 concludes Part III and the book by stressing on the detection of intrusions with data streaming concept.

Göteborg, Sweden
January 2020

Ismail Butun

Acknowledgements

First of all, my special thanks go to Prof. Ian F. Akyildiz (Ken Byers Chair Professor in Telecommunications, School of Electrical and Computer Engineering, Georgia Institute of Technology, Atlanta, Georgia, USA) for reviewing my book and providing his valuable feedback along with the *Foreword* section he has written.

I would like to express gratitude to my colleagues (especially to Magnus, Marina, Tomas, and Vincenzo) at Network and Systems Division, Department of Computer Science and Engineering, Chalmers University of Technology, for their courage and support.

Generous academicians and practitioners helped in the thorough review process of the book, including the authors of each chapter of this book. I appreciate each of them for providing their expertise in this process along with their valuable time.

Especially, I would like to convey my gratitude to the following external reviewers:

- Daniel dos Santos (Ph.D.), Forescout Technologies Inc. (USA)
- Lakshmikanth Guntupalli (Ph.D.), Ericsson Inc. (Sweden)

Last but not least, I would also like to thank my editor Susan Evans (Springer Nature, USA) and her team for providing the editorial support needed while preparing this book.

Goteborg, Sweden
January 2020

Ismail Butun

About the Editor

Ismail Butun (Ph.D.) received his B.Sc. and M.Sc. degrees in Electrical and Electronics Engineering from Hacettepe University. He received his second M.Sc. degree and Ph.D. degree in Electrical Engineering from the University of South Florida in 2009 and 2013, respectively. He worked as an Assistant Professor in years between 2015 and 2017 at Bursa Technical University and Abdullah Gul University. From 2016 to 2019, he was employed as a post-doctoral researcher by the University of Delaware and Mid Sweden University, respectively. Since July 2019, he has been working as a post-doctoral fellow for Network and Systems Division, Department of Computer Science and Engineering at Chalmers University of Technology. He has more than 36 publications in international peer-reviewed scientific journals and conference proceedings, along with an H-index of 12 and I-index of 14.

Dr. Butun is a well-recognized academic reviewer by IEEE, ACM, and Springer, who served for 39 various scientific journals and conferences in the review process of more than 106 articles. He contributed as a track chair and session chair for numerous international conferences and workshops, and performed as a technical program committee (TPC) member for several international conferences organized by IEEE, Springer, and ACM. His research interests include but not limited to computer networks, wireless communications, WSNs, IoT, IIoT, LPWAN, LoRa, cyber-physical systems, cryptography, network security, and intrusion detection.

Contents

Contributors

The authors are listed according to the appearances of the chapters they have written.

Alexios Lekidis Aristotle University of Thessaloniki, Thessaloniki, Greece

Alparslan Sari Cybersecurity Research Group, Department of Electrical and Computer Engineering, University of Delaware Newark, DE, USA

Hasan Basri Celebi Electrical Engineering and Computer Science Department, KTH Royal Institute of Technology, Stockholm, Sweden

Antonios Pitarokoilis Electrical Engineering and Computer Science Department, KTH Royal Institute of Technology, Stockholm, Sweden

Mikael Skoglund Electrical Engineering and Computer Science Department, KTH Royal Institute of Technology, Stockholm, Sweden

Daniel Minoli DVI Communications, New York, NY, USA

Benedict Occhiogrosso DVI Communications, New York, NY, USA

David Camacho Castillón GEA, Alcobendas, Madrid, Spain

Jorge Chavero Martín GEA, Alcobendas, Madrid, Spain

Damaso Perez-Moneo Suarez GEA, Alcobendas, Madrid, Spain

Álvaro Raimúndez Martínez GEA, Alcobendas, Madrid, Spain

Victor López Álvarez Telefónica I+D, Ronda de la Comunicación S/N Madrid, Madrid, Spain

Monjur Ahmed Waikato Institute of Technology, Hamilton, New Zealand

Sapna Jaidka Waikato Institute of Technology, Hamilton, New Zealand

Nurul I. Sarkar Auckland University of Technology, Auckland, New Zealand

Ismail Butun Network and Systems Division, Department of Computer Science and Engineering, Chalmers University of Technology, Göteborg, Sweden

Magnus Almgren Network and Systems Division, Department of Computer Science and Engineering, Chalmers University of Technology, Göteborg, Sweden

Vincenzo Gulisano Network and Systems Division, Department of Computer Science and Engineering, Chalmers University of Technology, Göteborg, Sweden

Marina Papatriantafilou Network and Systems Division, Department of Computer Science and Engineering, Chalmers University of Technology, Göteborg, Sweden

Acronyms

AI	Artificial Intelligence
AMI	Advanced Metering Infrastructure
AWGN	Additive white Gaussian noise
BACnet	Building Automation and Control Networks
BAS	Building Automation System
BBU	Baseband Unit
BCH	Bose, Chaudhuri, and Hocquenghem code
Bi-AWGN	Binary input AWGN channel
BLE	Bluetooth Low Energy
BMS	Building Management Systems
BYOD	Bring-Your-Own-Device
CBA	Component Based Automation
CER	Codeword Error Rate
COB	Communication Objects
COSEM	COmpanion Specification for Energy Metering
COTS	Commercial off the shelf
CPS	Cyber-Physical Systems
CSF	Cybersecurity Framework
CSMS	Cyber Security Management System
CSP	Cloud Service Provider
CV	Computer Vision
DB	Database
DBMS	Database Management System
DLMS	Device Language Message specification
DoS	Denial-Of-Service
DDoS	Distributed DoS
DIDS	Distributed IDS
DNP3	Distributed Network Protocol-3
EDS	Electronic Data Sheet
eMBB	enhanced Mobile Broadband
EPL	Ethernet Power-Link

EtherCAT	Ethernet for Control Automation Technology
GDPR	General Data Protection Regulation by EU
HART	Highway Addressable Remote Transducer protocol
HMI	Human–Machine Interface
IaaS	Infrastructure as a Service
IACS	Industrial and Control System
ICT	Information and Communications Technology
IDS	Intrusion Detection Systems
IEC	International Electrotechnical Commission
IIC	Industrial Internet Consortium
IIoT	Industrial Internet of Things
IoT	Internet of Things
IIRA	Internet Reference Architecture
IPS	Intrusion Prevention System
IRT	Isochronous Real Time
ISA	International Society of Automation
ISO	International Organization for Standardization
IT	Information Technology
LAN	Local Area Network
LDPC	Low Density Parity Check code
LED	Light Emitting Diode
LoRa	Long-Range
LoRaWAN	LoRa Wide Area Network
LPWA	Low-Power Wide-Area
LPWAN	Low-Power Wide-Area Networks
LTE	Long-Term Evolution
MAC	Medium Access Control
MITM	Man-In-The-Middle
ML	Machine Learning
ML	Maximum-likelihood
mMTC	massive Machine-Type Communication
MOS	Metal-Oxide Semiconductor
M2M	Machine to Machine
NB-IoT	Narrow-Band IoT
NERC	North American Electric Reliability
NFC	Near Field Communication
NIST	National Institute of Standards and Technology
NLP	Natural Language Processing
NMT	Network Management
OPC-UA	Object Linking and Embedding for Process Control Unified Architecture
OS	Ordered Statistics
OSI	Open System Interconnect
OT	Operational Technology
RH	Radio Head

RLC	Radio Link Control
PaaS	Platform as a Service
PAN	Personal Area Network
PDCP	Packet Data Convergence Protocol
PDN	Packet Data Network
PDO	Process Data Object
PERA	Purdue Enterprise Reference Architecture
PII	Personal Identification Information
PIIoT	Private IIoT
PLC	Programmable Logic Controller
PLM	Product Life-Cycle Management
PoE	Power over Ethernet
QoS	Quality of Service
RAMI4.0	Reference Architecture Model Industrie 4.0
RAT	Remote Access Trojan
ROI	Return of Investment
RTU	Remote Terminal Unit
SaaS	Software as a Service
SCADA	Supervisory Control And Data Acquisition
SCNM	Slot Communication Network Management
SDN	Software-Defined Networking
SDO	Service Data Object
SDR	Software-Defined Radio
SIS	Safety Instrument Systems
SNR	Signal-to-Noise Ratio
SPE	Stream Processing Engine
SVM	Support Vector Machine
TCP	Transmission Control Protocol
TMR	Triple Modular Redundancy
UAV	Unmanned Aerial Vehicles
UDP	User Datagram Protocol
UE	User Equipment
URLLC	Ultra-Reliable Low-Latency Communication
VANET	Vehicular Ad-hoc NETworks
VPN	Virtual Private Network
WiFi	Wireless Fidelity
WirelessHART	Wireless version of the wired HART protocol

Part I
Preliminaries, Design Principles and Challenges

Chapter 1
Industrial Networks and IIoT: Now and Future Trends

Alparslan Sari, Alexios Lekidis, and Ismail Butun

1.1 Introduction

Industry 4.0 revolution can be summarized with one word: 'Connectivity'. Connectivity will enable intelligent production with the proliferation of IIoT, cloud and big data. Smart devices can collect various data about indoor location, outdoor position, status information, usage patterns of the clients, etc. They have the ability not only in gathering information, but also sharing the information amongst intended peers. This will be beneficial in building an efficient manufacturing process in industrial environments and also in helping with the planned preventative maintenance on machinery. The other benefit is in identifying errors in the production pipeline as quickly as possible since it is an important factor to reduce the production and maintenance costs. Industry 4.0 is also focusing on optimization problems in the industry by using smart devices to utilize data-driven services. Industry 4.0 and IIoT are used for complex task sharing, decision making based on collected data, and remote access to machinery. Massive connectivity of the things and data collection/sharing capability of those promotes security to be a major requirement for the IIoT and Industry 4.0 concepts.

A. Sari (✉)
Cybersecurity Research Group, Department of Electrical and Computer Engineering, University of Delaware, Newark, DE, USA
e-mail: asari@udel.edu

A. Lekidis
Aristotle University of Thessaloniki, Thessaloniki, Greece
e-mail: alekidis@csd.auth.gr

I. Butun
Network and Systems Division, Department of Computer Science and Engineering, Chalmers University of Technology, Göteborg, Sweden
e-mail: ismail.butun@chalmers.se

I. Butun (ed.), *Industrial IoT*, https://doi.org/10.1007/978-3-030-42500-5_1

Semiconductor transistors are introduced in the late 1940s and led the microprocessor revolution in the 1970s. Intel produced MOS (Metal-Oxide Semiconductor) based 4 bits 4004 microprocessor in 1971. It had 2250 transistors, 10,000 nm MOS process and area of 12 mm^2. Figure 1.1[1] shows the plot of MOS transistor counts by years. In the plots, the exponential increase of transistor counts validates Moore's law—transistors in an integrated circuit doubles every 2 years. Currently, transistor counts jumped to 9–40 billion with 7–12 nm MOS process and area of 100–1000 mm^2. Figures 1.2 and 1.3 shows the evolution of the flash memory and RAM with transistor count vs. date plot. Groundbreaking technological advances in electronics started the information age which triggered major changes in communication technologies such as the Internet (connectivity), advanced machinery, and software development, etc. Internet became the global communication hub, enabling us to exchange information instantly. Advanced electronics produced smart devices with a smaller size, which constitute today's 'things' of IoT and IIoT.

IoT is the proliferation of smart devices such as tablets, phones, home appliances such as TVs, and other sensors, etc. The benefit of using smart devices at home would be reducing electric bills and time savings etc. Managing resource usage based on sensors or scheduling heavy-duty tasks like running dishwasher, washing machine or dryer when the electric consumption is the cheapest. IoT devices are commonly used by the hobbyist or another consumer usage, and even in industry. However, IIoT is designed for heavy-duty tasks such as manufacturing, monitoring, etc. So, IIoT uses more precise and durable (heat/cold resistant) devices, actuators, sensors, etc. Both IoT and IIoT have the same core principles such as data management, network, security, cloud, etc. The main differences between IoT and IIoT are scalability and the volume of generated data and how data has been handled. Since IIoT devices generate massive amount of data, IIoT requires data streaming, big data, machine learning or artificial intelligence practices. In a home network, loss of the generated data would be trivial but in IIoT it is vital. The data in IIoT should be more precise, continuous and sensitive. For instance, considering a monitoring system in a nuclear power plant or a manufacturing facility should be precise, continuous and sensitive to prevent hazardous events. The implementation of IIoT in production lines or other industrial projects, companies are aiming to reduce production or maintenance costs and improve efficiency, stability, safety, etc.

According to a key note speech[2] delivered by Tom Bradicich,[3] the seven principles of the IIoT are provided as follows:

- *Big amount of analog data:* Many sensors generate analog data and this data needs to be digitized to be further treated, analyzed and stored.
- *Perpetual connectivity:* Devices of the IIoT are always connected. There are three key benefits of this: (1) Real-time monitoring is possible. (2) Continuous

[1]Figures 1.1, 1.2, and 1.3 are illustrated based on data from: https://www.wikiwand.com/en/Transistor_count.

[2]Available at: https://www.youtube.com/watch?v=u3IaXvjDiOE.

[3]Tom Bradicich, Ph.D., VP and GM, Servers and IoT Systems, Hewlett Packard Enterprise.

Fig. 1.1 Figure shows transistor counts in microprocessors by date between (**a**) 1971–1990 (top), (**b**) 1991–2010 (middle), and (**c**) 2010–2019 (bottom)

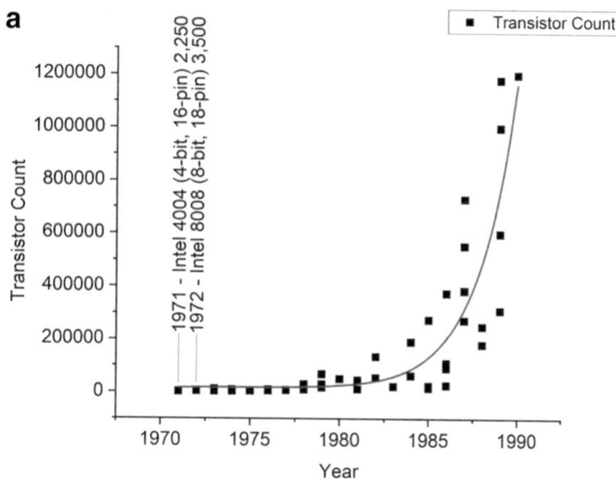

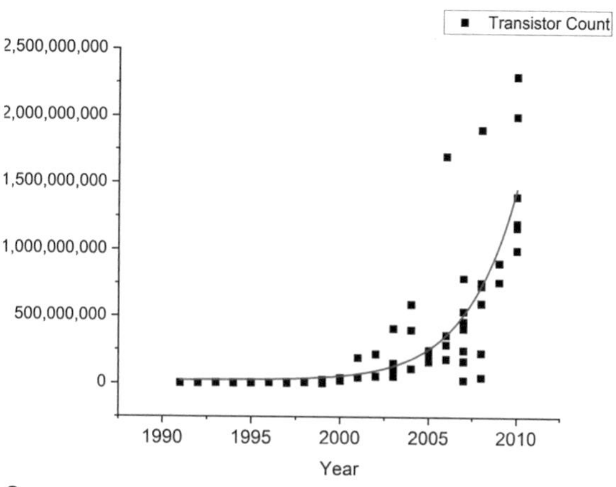

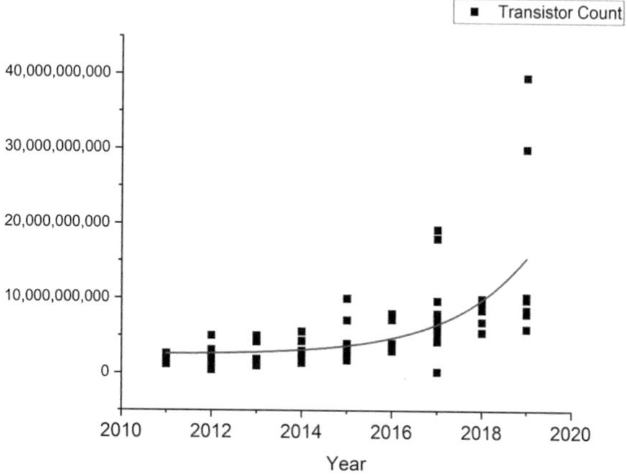

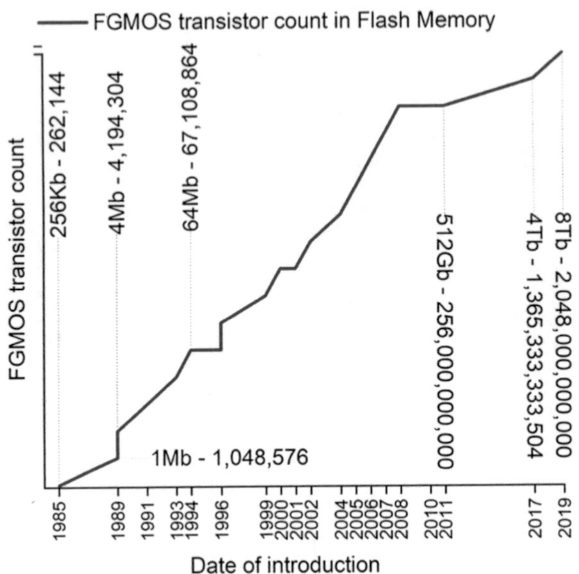

Fig. 1.2 FGMOS transistor count in Flash memory over years with capacity increase

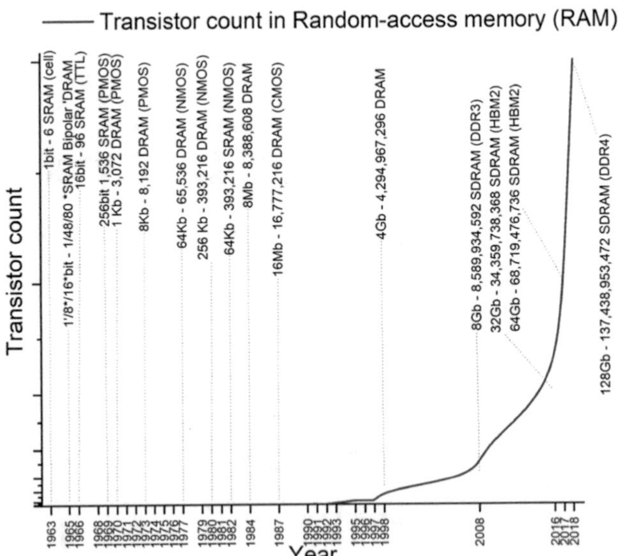

Fig. 1.3 Transistor count in RAM over years with capacity increase

monitoring can help us to push software-firmware updates and fixes. (3) The benefit of connected devices motivates individuals and organizations to purchase products.

- *Real-time data streaming:* In the industry there are many safety mechanisms are in use, and they are constantly generating data. Considering a nuclear power plant, safety is of utmost critical from the operations point of view. Monitoring requires real-time data streaming since a possible delay in data would cause disastrous events. Hence, real-time data streaming and its aggregation are really important.
- *Data insights:* Data insights (Spectrum of Value) in IoT seeks an answer to the following question: "What are you trying to achieve?"
- *Time-to vs. depth-of -insight trade-off:* It is equivalent to the immediacy-of-knowledge compared to the depth-of-knowledge. For instance, while monitoring or analyzing a nuclear power plant data, immediate attention is required, whereas an analysis of a scientific experiment data (data by CERN or NASA) can take years to uncover scientific problems.
- *Visibility from Big Data:* Once the data is collected and stored in big data environment, later on it should be available whenever it is needed for analysis or other tasks.
- *Edge computing:* Data center class computing and analytics will be shifted to edge (latency, bandwidth, cost, security, duplication, reliability, corruption, compliance, and data sovereignty).

Improvement in big data and data streaming technologies enabled organizations to use Artificial Intelligence (AI) and Machine Learning (ML) products more widely in Industry 4.0. AI and ML applications are emerging in health, education, defense, security, industry, etc. Many technology giants are pouring billions of dollars to develop AI products such as autonomous cars (self-driving cars). Google, NVIDIA, and others are developing computer vision-based self-driving algorithms along with pedestrian detection, collision detection, etc. After the famous Urban Challenge organized by DARPA, autonomous cars brought to the spotlight once again. Automobile companies like Ford, General Motors, Nissan, Tesla, Mercedes, etc. invested billions in R&D. Hundreds of other small companies are building Radars, cameras, computing and communication systems, other sensors, etc. According to the NVIDIA developers' blog, autonomous cars are responsible for the generation of an enormous amount of data involved with the following equipment: Radar, Sonar, GPS, Lidar and cameras. A single forward-facing Radar (2800 MBits/s) generates approximately 1.26 TB of data per hour. A two-megapixel camera (24 bits per pixel) operating at 30 frames per second generates 1440 Mbits of data per second (approximately 1TB data per hour [29].

In computer science, the following AI disciplines are utilized for various purposes: Natural Language Processing (NLP), multi-agent systems (coordination and collaboration—distributed resolution of problems, decision and reasoning, learning, planning, simulation), human interactions (learning, chat-bots, expert systems), computer vision, robotics, neuroscience and cognitive science (comprehension

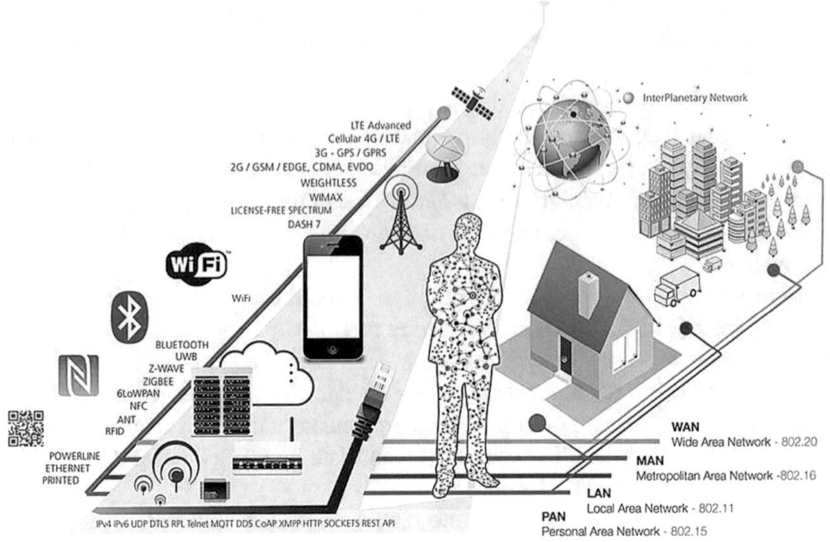

Fig. 1.4 IoT connectivity diagram (see footnote 4)

and simulation of the brain and nervous system), decision support (game theory, uncertainty, explicability). It is assumed that using AI and ML algorithms will contribute efficiency, cost reduction and effective management of industrial networks. Therefore, following AI-based algorithms are also utilized frequently: heuristics, logical programming, deduction and proof, reasoning, planning, scheduling, and search.

The wired/wireless communication technologies that are being used for IIoT networks are as shown in Figs. 1.4[4] and 1.5. They are also summarized here as follows:

BLE Bluetooth Low Energy (BLE) is a wireless network technology of Personal Area Networking (PAN), which is widely used in smart devices such as phones, watches, wearable electronics, etc. BLE applications can be found in smart home, health and sport industry. Many operating systems have native support for this technology such as Android, BlackBerry, IOS, Linux, macOS, Windows, etc. BLE uses 2.4 GHz radio frequencies.

Wi-Fi Wireless Fidelity (Wi-Fi) is an IEEE 802.11 based wireless network technology that is widely used with computers, tablets, smart devices, and vehicles such as cars, buses, drones, etc.

[4]Source: Postscapes and Harbor Research, available at: http://postscapes.com/what-exactly-is-the-internet-of-things-infographic/.

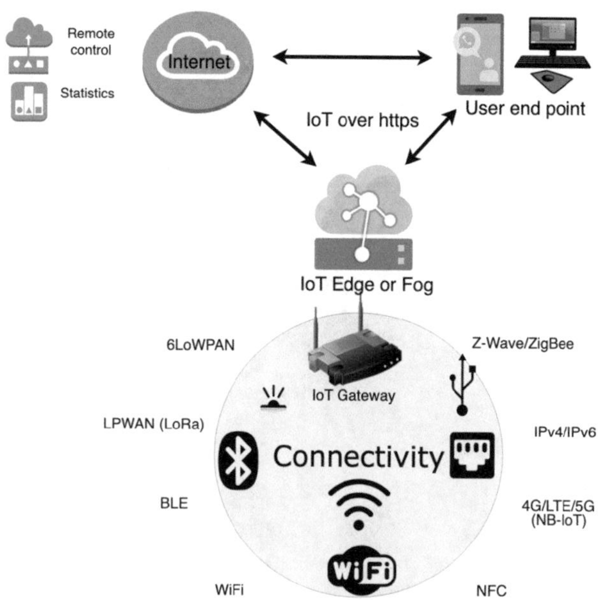

Fig. 1.5 A subset of IoT and IIoT communication technologies and protocols are illustrated

IPv4/IPv6 Internet Protocol versions 4/6 (IPv4/IPv6) can be considered as the backbone of the Internet and other packet-switched based networks. It is one of the most widely used technology in real life.

NFC Near Field Communication (NFC) is a communication technology which devices can do connection approximately within 4 cm distance. NFC devices are commonly used in electronic payments, key-cards, electronic tickets, etc. Many smartphones are supporting this technology such as Android, Blackberry, IOS, etc.

Z-Wave/ZigBee Z-Wave is a low energy wireless protocol. ZigBee is also a low power (low energy consumption) wireless communication protocol based on IEEE 802.15.4 standard to create PAN. Z-Wave/ZigBee are commonly used in home automation, medical and industrial applications.

4G/LTE/5G 4G (fourth generation), LTE (Long Term Evolution), and 5G (fifth generation) are dominating broadband cellular network technologies in nowadays mobile telecommunications.

NB-IoT Recently, NB-IoT (Narrow Band Internet of Things) rapidly deployed by mobile operators with the proliferation of IoT. NB-IoT is an LPWAN standard that focuses on specifically low cost, energy-efficient, and indoor coverage.

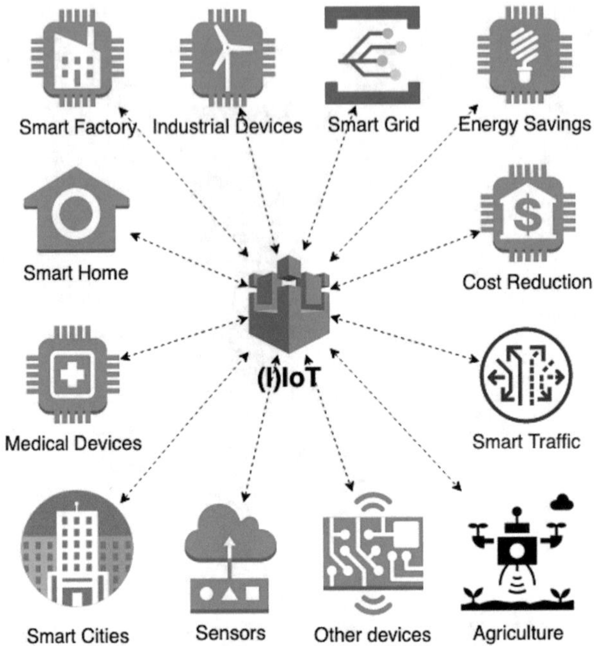

Fig. 1.6 A subset of IoT and IIoT use cases are illustrated

According to one of the leading sector representative[5] of industrial investment companies, IIoT will constitute one of the main pillars of Industry 4.0 and would be described as follows: "The IIoT is a network of physical objects, systems platforms and applications that contain embedded technology to communicate and share intelligence with each other, the external environment and with people." By following this definition and the technological trends, economic predictors[6] deduct this conclusion: "The IIoT has the capacity to significantly boost the productivity and competitiveness of industrial economies, but poor supporting conditions - especially the lack of digital literacy - will hold many countries back".

There will be plenty of application areas for IIoT ranging from smart cities to precision agriculture, smart traffic to smart grid (the future of electric grid), and so on, as shown in Fig. 1.6.

As discussed thoroughly in [11], IoT networks are vulnerable to various types of attacks and weaknesses. Due to the absence of a physical line-of-defense, security of such networks is of a big concern to the scientific community. Besides, as cyber-attacks on industrial environments happen more frequently and their results

[5]More information is available at: https://www.accenture.com/us-en/.

[6]More information is available at: http://www.businesswire.com/news/home/20150120005160/en/\Industrial-Internet-Boost-Economic-Growth-Greater-Government.

are elevated to a devastating scale, effective intrusion detection is of the utmost importance of industrial networks. Hence, research regarding Intrusion Detection Systems (IDSs) for industrial networks is arising.

Prevention and early detection of intrusions are important, especially for mission-critical industrial networks, such as smart grid systems.

For instance, data streaming concept provided in later chapters (Chap. 6), might indeed enhance detection/response time in IDSs of the industrial networks. Hence, future IIoT implementations can benefit from data streaming not only to reduce up-streaming data size, but also to improve the security of the network especially by enhancing the overall IDS response time.

Nowadays, there exist example applications in industrial networks (such as smart grid systems), where sensors can generate data at rates of ∼3 GB (Gigabyte)'s per hour [46, 54]. This is way much beyond the Ericsson's and Bloomberg's expectation [38], which was ∼280 PB (Petabyte) of data to be generated by 680 million smart sensors per year (which is equivalent to 1 byte of information per sensor per year). However, current client/server architecture of the traditional networks is not capable of transferring, storing, and processing such big volume of data. Hence, a sampled version of the generated data is used. Secondary benefit of data streaming would be utilization of the unused portioned of the mentioned big data from which IDSs would also benefit.

Centralized system architectures that gather all the data in the cloud to process it cannot scale, both due to communication bandwidth and issues of scalability of processing accumulated data; e.g. matching patterns of data streams, on-the-fly, is a lower-latency process compared to searching for patterns in terabytes/petabytes of stored data. Due to this, only small fractions of generated data is actually being processed and used in such infrastructures. With this reasoning, it is possible to see that besides helping resolving bottleneck situations, stream processing can enable usage of more data for applications in general and hence for IDSs as well.

So far, a brief overview of industrial networks and IIoT is presented in Sect. 1.1. The rest of the chapter is structured as follows: Challenges faced by the practitioners and the researchers while working on industrial networks are presented in Sect. 1.2. Section 1.3 projects the future of industrial networks, including the technologies that can be adopted by. Whereas, Sect. 1.4 presents enabling technologies for industrial networks. Finally, Sect. 1.5 concludes the chapter.

1.2 Challenges in Industrial Networks

Industrial networks are subject to many technical challenges that need to be considered before, during, or after any implementation. These challenges are as follows but not limited to:

1.2.1 Wireless Coexistence

Wireless coexistence stands for the safe operation of the devices that are using wireless technology for the purpose of their communication. Especially, 'signal interference' is the main problem against coexistence of different wireless communication standards, which might cause fading of the signals if they are operating in the same frequency bands. This is on top of the existing wireless communication problems such as reflection, refraction, diffraction, scattering and *free space path loss* of the signals on air.[7] Eventually, wireless signal interference might cause packet drop, data loss, jitter and delay in transmission, and an asynchronization between the two communicating parties. Therefore, while designing IIoT, wireless coexistence of the devices should be maintained, for instance by providing enough physical distance in between the devices, and or by efficiently dividing and sharing the frequency spectrum that is being used.

1.2.2 Latency

Latency, sometimes referred to as delay, (for the industrial networks) is the time passed between the release time of a specific command and the start time of the execution for that specific command. In some specific cases, it might be also referred to as the time passed in between the data collection and the output of reaction. For instance, even with the existence of cloud-end implementation, Ferrari et al. have shown that, by using inexpensive industrial grade IIoT devices, the round-trip latency of IIoT applications can be less than 300 ms for inter-continental communications and less than 50 ms for intra-continental communications [28]. Although, any figure below 300 ms is considered as sufficient for telephony communication to avoid undesired "talk-over" in conversations, every ms counts when the industrial networks are considered. Especially real-time response might be needed for the installments that are involving high speed machinery with safety requirements.

[7] More reading available at: https://www.accoladewireless.com/wlan-wifi-signal-issues/.

These important aspects *(wireless coexistence* and *latency*) of wireless communication, which is the enabling technology for IIoT and IoT networks, are thoroughly discussed in Chap. 2 of this book. Especially, next generation wireless technologies, such as 5G and the ultra-reliable low-latency communication (URLCC) are presented including the theoretical limits and the *latency*. The trade-offs in low-latency communication for receivers with computational complexity constraints are carefully discussed. Accordingly, it is stressed that, for mission critical IIoT deployments such as the factory implementations requiring real-time response, the latency is utmost important. With very stringent latency requirements and low-complex IIoT receivers, the time required for the decoding of a packet (should not constitute a computationally demanding operation) must be also considered while analyzing the total latency.

1.2.3 Interoperability

Interoperability refers to the essential capability of various computerized merchandises or systems to promptly connect and exchange data with each other, without facing any restriction. There are two types of interoperability issues defined for IIoT [31]:

1. The cross-layer interoperability (also known as heterogeneity), which is defined as the orchestration of the Open System Interconnect (OSI) layers in a seamless and burden-less way. Recently, this issue is solved by using two emerging virtualization technologies; namely, Software-Defined Radio (SDR) and Software-Defined Networking (SDN) [5]. On one hand, SDN is a network architecture based on the separation of the control plane and the data plane which makes it possible to obtain a directly programmable network. This separation unifies the control plane over all kinds of network devices, making the network configured and optimized based on the network status [26]. On the other hand, by providing a higher degree of flexibility in designing wireless interfaces competent enough to support multiple communications technologies, SDR [39] can be a solution for addressing the interoperability problems. Akyildiz et al. [52] mentioned that a vast number of devices in IoTs constitute a fundamental challenge to the ubiquitous information transmissions through the backbone networks. The heterogeneity of IoT devices and the hardware-based, inflexible cellular architectures possess even greater challenges to enable efficient communication. Hence, they offered an architecture on wireless SDN and proposed software-defined gateways that jointly optimize cross-layer communication functionalities between heterogeneous IoT devices and cellular systems.
2. The cross-system interoperability, which is defined as maintaining the continuous operation of the networks and the systems within the existence of different system architectures. E.g., part of an IIoT might consist of LoRa-based end-devices and the other part might be dominated by Nb-IoT end-devices. This

issue is may be solved by using semantic system models that provide a deeper understanding of the raw sensor data, by enabling AI-based machines to make decisions based on simple rules [23]. The design of gateways for interoperability can also be achieved by defining comprehensive centralized metadata and defining abstract models out of them that can be automatically generated for multiple programming languages.

1.2.4 Sensor Data Streaming

Data streaming is a novel and powerful tool for industrial networks, which will enhance capabilities of the data analysis teams (especially in cyber-security). However, there are several challenges that needs to be accounted for [1]:

Plan for Scalability Seeks an answer for how much granularity and processing is needed on the data so that the network scales reasonably and can be kept under control.

Plan for Data Durability Seeks ways of long-term data protection while storing the data, so that they do not suffer from bit fault, degradation or other corruption. Rather than focusing on hardware redundancy, durability is more concerned with data redundancy so that data is never lost or compromised.

Plan for Fault Tolerance Incorporating fault tolerance in both the storage and processing layers is also important.

1.2.5 Safety

IIoT sensors and devices play critical a role in industrial (manufacturing, transportation, etc.) safety. IIoT leverages low cost and low power devices to implement energy-efficient safety protocols while improving productivity. The primary focus of safety is on internal risk-based problems in the manufacturing pipeline to prevent broad-spectrum issues like everyday work related to small accidents and as well as major disasters such as nuclear accidents and incidents. Many industrial facilities utilize multi-layer safety protocols,[8] such as process control monitoring and alarms, Safety Instrument Systems (SIS), physical protection systems, emergency response (local/external) systems.

[8]More information available at: https://www.isa.org/intech/201804web/.

Alarms constitute a mechanism to inform a certain threshold is breached for a monitored event. *Monitoring* (see footnote 8) is one of the most critical components of safety protocols. Most of the production pipelines require real-time continuous monitoring. Sensors generate real-time analog/digital data and transmit them to a control unit to analyze and perform activities based on the monitoring value. Latency is an important problem in monitoring systems. Moreover, physical monitoring is also important like corrosion/erosion monitoring to figure out detrimental effects. Physical protection system involves following equipment: Pressure-relief valves, rupture discs, stream traps, electrical switch-gear, eyewash stations, safety showers, etc.

Recent trends in IIoT safety systems are as follows:

- *Computer Vision (CV):* The CV technology is being used for anomaly detection (in detecting physical structural or machinery problems) along with other IIoT sensors.
- *Wearable electronics:* Involves monitoring of Vital life-signal measurements, fall detection, electric shock detection, etc.
- *IIoT security to ensure safety:* Cybersecurity is also a major concern for safety systems. All IIoT devices should be cybersecurity proof during the operational phase. An external party like competitors or hackers should not halt or cause physical destruction of a production facility or a pipeline. For instance, in June 2010, Iran suffered a cyber-attack in a nuclear facility located at Natanz via Stuxnet cyber-worm [27]. Another major incident was on 31 March 2015, Iranian hackers took down the Turkish power grid which caused a massive power outage of 12 h in 44 out of 81 cities. Approximately 40 million people were affected during this outage [30].
- *Robotics:* AI-driven software technologies are embedded to achieve autonomous decision-making machinery called robots. These devices can operate where a human can not. Such as in nuclear power plant or under the sea, etc.

1.2.6 Security and Privacy

Interfacing the smart factories with Cyber-Physical Systems (CPSs) and IIoT improves the intelligence of the infrastructures, yet introduces cyber-security vulnerabilities which may lead to critical problems such as system failures, privacy violations and/or data integrity breaches. As the privacy of the citizens is becoming into prominence, especially in the EU with the GDPR (General Data Protection Regulation—2016/679) act, privacy bearing information of IIoT users such as Personal Identifying Information (PII), need to be treated well, so that they will be kept confidential [7].

As thoroughly discussed in [11] and also Chap. 5 of this book, many IoT networks do not even possess basic security elements. On average, these are the cyber-security analysis of today's COTS (Commercial off the shelf) IoT products[9]: 25 vulnerabilities are detected per device, 60% has vulnerable firmware's and user interfaces, 70% do not encrypt any communications at all, and 80% fails to request password for authentication that has a secure length. Henceforth, there are two main methods to fight against intrusions and cyber-attacks against IIoT. One of them is allowing intrusions to happen and then detecting them via Intrusion Detection Systems (IDS) as in [3]. The other one is prevention of attacks by means of authentication, authorization and access control.

1.2.6.1 Intrusion Detection

In practice, IDSs are installed and on demand from every aspects of technological life, from corporate to universities where IT department exists. Especially, following are the topics of our interest:

- Industrial network and IIoT security
- Smart grid security
- Critical infrastructure security
- Smart city/factory/home security

The main distinction among the anomaly-based IDS and misuse/specification-based IDS is, anomaly detection-based systems can detect any kind of bad behavior (theoretically) on the fly but misuse/specification detection-based systems can only recognize previously known bad behaviors (signatures) [8].

The reasoning behind this is as follows: Misuse/specification detection-based systems are designed in a way to match previously modelled attack vectors and rules. If an incidence can be categorized in these, then it is called as an attack. Otherwise, it is called as normal behavior. These systems work very well with the defined and categorized attacks up until the time of implementation. However, they are quite useless in the case of new attack vectors that can not be specified with the old ones. For these situations, anomaly detection-based IDS is suggested. Hence it is easy to model the normal state than the abnormal one, it is modelled to specify the good (or normal) phase of the system behaviors. Therefore, it is logically opposite of the misuse/specification detection-based systems in which the attack signatures and vectors are directly identified. So, anomaly detection-based IDS uses a kind of live reasoning algorithm, meaning that the decision on a future incident might differ from a past one, even if they are the same events.

Besides, as mentioned deeply in above discussions, they are easier to setup, as modelling the normal operating conditions (phase) of the network is easier

[9]By COTS vendors such as ABB, Arm, Bosch, Huawei, Intel, Siemens, Netvox, etc.

than specifically identifying each attack vector. Therefore, the next subsection is dedicated for that.

Each of these approaches is then sub-classified into various methods. The interested reader is referred to [9] and also Chap. 6 of this book, for a more detailed discussion.

1.2.6.2 Intrusion Prevention

It is referred to as taking all necessary actions required in order to prevent intrusions. It might be analogically similar to theft protection systems in real-life security applications such as installment of advanced door locks, infrared detectors, etc. Some of the methods to be mentioned are as follows but not limited to [6]: Authentication, authorization, access control, ciphering (encryption/decryption), hashing, etc.

1.2.6.3 Runtime Security Monitoring

The runtime security monitor operates on the state of the industrial system and observes anomalous behavior that occurs either by failures and operational errors or security threats by adversaries. To guarantee its successful operation the runtime security monitor requires knowledge about the system as well as the trusted states, operation conditions as well as the system output information i.e. what the system produces. The objective of the runtime security monitor is to identify suspicious and anomalous indicators when i.e. the system stops producing what is intended. These are depicted as faults or security threats (Fig. 1.7). Then, it acts proactively by replacing the system with a so-called "reversionary" system, that is performing the minimum industrial system functionality defined by the requirements. The reversionary system ensures redundancy and reliability.

The presence of a runtime security monitor should be always coupled with an industrial network analyzer (Fig. 1.7), as the events and data that are exchanged in the industrial network are encoded in proprietary protocols and message formats. An

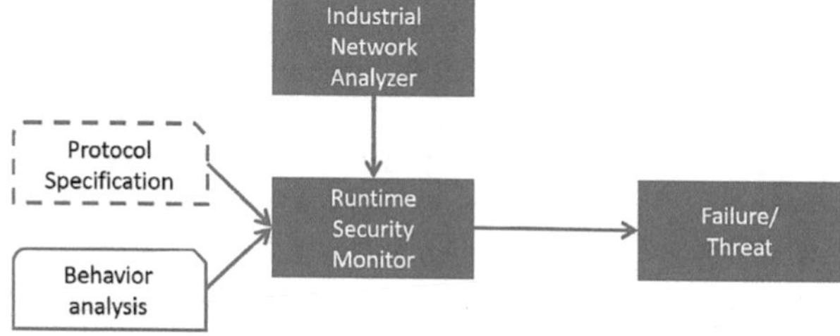

Fig. 1.7 The concept of runtime security monitoring

example of a network analyzer is Wireshark[10] providing support for some industrial protocols such as Modbus and DNP3 as well as libraries and extensions for other proprietary protocols (e.g. Siemens Step 7).

1.3 Future Trends in Industrial Networks

Here, it is worth mentioning that, WSN represents an earlier version of sensors (sensing devices) that have sensing, communication and networking capabilities to set up a network and send sensed information to the data sink throughout the network. IoT is evolved from WSN notion with the need of reaching those sensors and actuators via global network of Internet. IIoT is a subset of the IoT especially devised to support industrial automation; therefore possess most of the characteristics of generic IoT networks. Hence, the OSI network structure of an IIoT (and IoT) consists of five layers as described in [35]: Physical, Data-Link (MAC), Network, Transport, and Application. It should be noted that Session and Presentation layers of the traditional OSI network model are all considered in the Application layer of IIoT (and IoT).

It is expected the future industrial networks to include more wireless technologies such as the emerging ones 5G, LoRa, etc., in order to support the *automation* notion of Industry 4.0 by enabling remote command, control and monitoring of the sensors and actuators on the factory floor. It might be expected these wireless technologies to enable Internet connectivity to industrial networks, which would eventually cause both terms, industrial networks and IIoT to be federated under same term. Extensively usage of automation over the industrial domain will also necessitate an increased number of remote sensor and actuator installations, resulting in the creation of a vast amount of data, called big data. Besides, as most industrial networks are mission-critical, the security and safety of these networks will be important. Therefore, filtering information from the big data will be one of the most challenging aspects of the industrial networks, as a timely response is really critical in the industrial environments. One recent solution to this dilemma (on where to process the data) is, pushing cloud/server tasks/missions towards the edge of the network, from which the term edge/fog computing emerged. Due to CISCO, *fog computing* extends and complements the *cloud computing* with the concept of smart devices which can work on the edge of the network [13].

Authors of this chapter envision that the *data streaming* paradigm will be an enabling technology for the *fog computing* notion of the CISCO, and therefore play a critical role at the future industrial networks. Data streaming will help fog computing, by processing the data closer to the source, decreasing the amount of traffic created, and also reducing the overall response time for the queries. This will be especially beneficial in detecting intrusions. According to the authors' insight, in

[10]https://www.wireshark.org/.

the near future, the data streaming paradigm (described thoroughly in Chap. 6 of this book) along with a distributed architecture will be adopted by industrial networks.

1.3.1 Industry 4.0

The industrial revolution started with the mechanization (water/steam power) of the production pipelines. Steam engines are used in production and transportation. The second transformation has happened once the steam power is replaced with electricity. Breakthrough in transistor and electric circuits technology produced computers which triggered the third industrial revolution. Mass device connectivity nudges the fourth industrial revolution which is enhanced with smart devices and advanced artificial intelligence algorithms like machine learning. The main components of Industry 4.0 and IoT are as follows: Cloud, cyber-security, IoT, system integration, simulation, autonomous robots, Big Data, augmented reality, and additive manufacturing. The history of Industry 4.0 is well explained in Sect. 4.2. Briefly, the focus of Industry 4.0 is to do optimization in computerization and eliminate human involvement in the decision-making process to eliminate human errors and improve efficiency. This can be done with smart devices (monitoring) and automation.

1.3.2 Industrial Site Indoor and Outdoor Lighting

IIoT-based systems, also in conjunction with Light Emitting Diode (LED) luminaires, intelligent controls by using RF and Power over Ethernet (PoE) technology, is enabling significant advancements in next-generation building lighting systems for commercial and industrial buildings as well as street lighting. Especially, Chap. 3 of this book is dedicated for this specific application area (commercial and industrial building lighting and in street lighting) of IIoT networks to industrial domain. The technical, economic, and market aspects of this group of technologies are discussed and presented thoroughly in Chap. 3.

1.3.3 Smart Grid Systems

Smart grid systems not only consist of an electricity distribution network, but also a coupled information network on top. These systems mostly consist of smart meters, nowadays of which mostly composed of IoT devices, hence might be a good example of an IIoT network.

In most cases, bottlenecks occur in the upstream path of the networks. Intrusion detection solution (especially related to the data streaming) discussed in this book

(Chap. 6), aims at pushing the data analysis towards peripheral devices instead of gathering data centrally. This kind of solution is offered to provide a fast response by removing the necessity of the round-trip messages in between the smart meter and the server. On the other hand, data streaming is also beneficial to the overall network performance due to the decreased load in the aforementioned bottlenecks.

1.3.4 NarrowBand IoT (NB-IoT)

Apart from the 3GPP existing 4G and ongoing 5G standards, 3GPP is also providing standards for cellular Low Power Wide Area Network (LPWAN) communication mechanisms and functionalities targeted in low-end devices, such as the ones used in the IoT devices. A representative example in this category is *Narrowband-IoT (NB-IoT)* or often referred to as LTE-M2 [45].

The NB-IoT provides low energy consumption, small volumes of data and transmission over large distances or deep within buildings. It is based on release 13 of 3GPP and operates at even lower bandwidths (180 kHz/channel) and lower data rates (20 kbps) in the licensed LTE spectrum. Mobility is sacrificed in favor of better indoor coverage and support for larger number of devices. NB-IoT is managed by cellular operators with expected costs and regulations on access to this network.

The 3GPP offers three scenarios for LPWAN deployment in NB-IoT (illustrated in Fig. 1.8), which are namely In-Band, Guard-Band and Standalone. Specifically, In Band makes use of the same resource block in the LTE carrier of the existing LTE network. Furthermore, guard-band deployment uses the unused blocks within the LTE carrier guard band and standalone deployment utilizes new bandwidth in comparison to existing technologies (e.g. GSM, LTE).

1.3.5 LoRa and LoRaWAN

(continued)

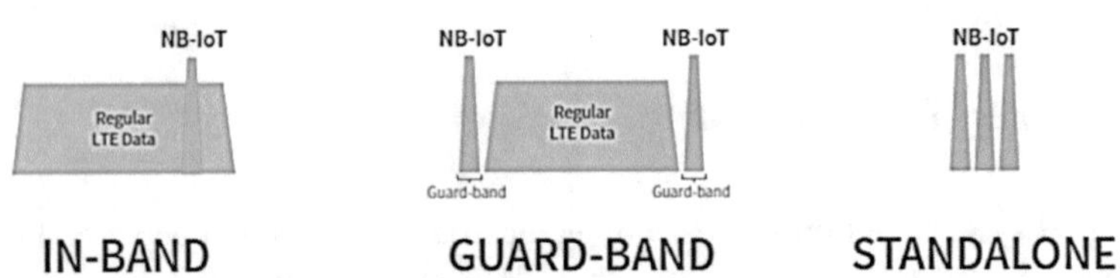

Fig. 1.8 Scenarios for LPWAN deployment in NB-IoT

LoRa (Long-Range) is a subclass of LPWAN or Low-Power Wide-Area (LPWA), which is a new class of communications technology devised for IoT network implementations. LoRa describes physical layer communications technology, which is a Chirp Spread Spectrum (CSS) to spread the signal in the frequency band in order to get resistance against wireless signal interference and fading [12]. LoRaWAN describes all upper layer OSI stacks to enable seamless packet transmission. Apart from its earlier version (v1.0), new version (v1.1) of LoRAWAN is known to be secure against most of the cyber-attacks as shown in [10] and [25].

LoRaWAN [37] is an openly defined network protocol that manages communication between gateways and end-devices with the following features: (1) establishing encryption keys for application payloads and network traffic, (2) device to gateway pairing assignments, and (3) channel, power and data rate selection. The devices in LoRaWAN can be of three types:

1. bi-directional end-devices with downlink followed by uplink, as for example sensor end devices,
2. bi-directional end-devices with transmission slots scheduled for downlink, as in the case of actuators, and
3. always-on bi-directional devices, which is intended for low-resource devices to ensure low-latency such as gateways or servers.

A reference view of the LoRaWAN architecture is provided in Fig. 1.9. The figure illustrates the communication between the LoRaWAN Server, the gateways as well as the end devices. The gateways are responsible for maintaining radio connectivity as well as may act as transparent bridge on the network. Furthermore, they ensure seamless network upgrade. Additionally, the LoRaWAN Server is responsible for maintaining association with end node, configuring data rates, removing duplicates and the handling security and access control interfaces with applications. Finally, the LoRaWAN end device in the system has a network communication and application encryption key. All packets are transparently sent from gateways to a LoRaWAN server without any local decryption to limit the potential risk of compromised clients and gateways (Fig. 1.9).

1.3.6 LTE (Long-Term Evolution) for Machine Type Communications

LTE-MTC or LTE-M is an LPWA technology standard based on 3GPP's Release 13 specification. It specifically refers to LTE Cat M1, suitable for the IoT. Even though both NB-IoT and LTE-M use LTE and aims in enabling low-power communication

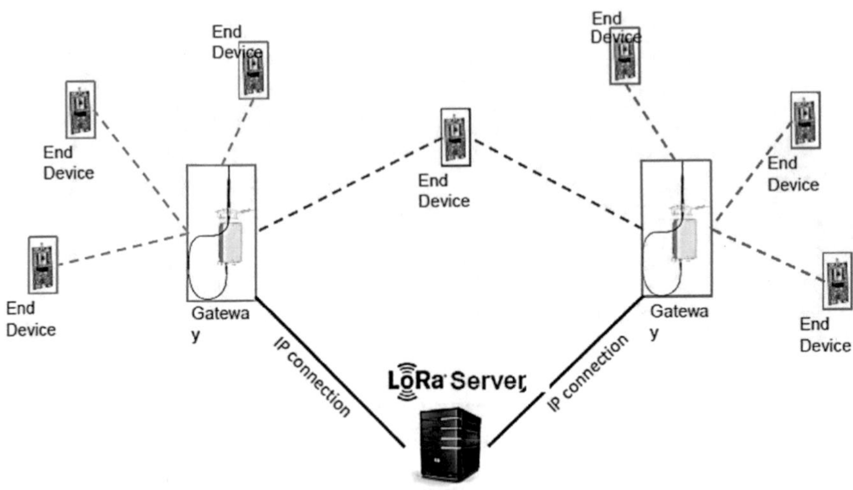

Fig. 1.9 Reference architecture in LoRa

Table 1.1 NB-IoT and LTE-M differences

Technology characteristics	NB-IoT	LTE-M
Bandwidth	<250 kbps (Half duplex)	384 kbps–1 Mbps (Half or full duplex)
Coverage	20 dB	15 dB
Mobility	No	Yes
Designed for	Message-based communication	IP-based communication
Power saving	eDRX [34]	Deep-sleep/periodic wake-up

in IoT devices, their main differences are in terms of throughput, mobility, power, latency, and cost.

Table 1.1 provides insight on the difference in technology characteristics for NB-IoT and LTE-M. The main differences are found in terms of mobility as well as the technology design. Specifically, NB-IoT does not provide mobility when change from one cell to another and the User Equipment (UE) have to perform idle rejoin. This introduces a power penalty for moving devices). NB-IoT is good for sending small and subsequent messages, whereas LTE-M is used to send sequences of messages, such as data streams. Additionally, LTE-M and NB-IoT have also a difference in the power saving mode that the support. In particular, LTE-M supports several power saving modes (e.g.: deep-sleep or wake-up only periodically while connected), whereas NB-IoT supports the Extended Discontinuous Reception (eDRX) [34] mode, which allows reduced power consumption for devices that are awake and remain connected.

1.3.7 5G

Wireless cellular technology in mobile communication has reached the 5G (Fifth Generation—2019) milestone. 1G (First generation) cellular networks designed with analog technology in 1979. Whereas, 2G (Second generation—1991, CDMA (Code-division multiple access), GSM (Global System for Mobile Communications), TDMA(time division multiple access)) cellular networks are digital which supports messaging like SMS and voicemail. 3G (Third generation—2001, EVDO (Evolution-Data Optimized), HSPA (Evolved High-Speed Packet Access), UMTS (Universal Mobile Telecommunications System)) has faster data transfer rate ~144 kbit/s. which enables GPS and mobile web technologies. 4G (Fourth generation—2009, WiMAX (Worldwide Interoperability for Microwave Access), LTE (Long-Term Evolution)) broadband utilizes IP and packet-switched technology to handle data rates of ~100 Mbit/s. 4G can be considered as the rise of the mobile Internet. 4G cellular networks can support the following applications: IP telephony, video conference, and streaming. 5G is the latest wireless technology that provides the fastest connectivity (download and upload speed). It has also following advantages: reliable connection, high-quality voice and video transmission, along with a support of increased number of connected (IoT) devices. 5G networks can be configured with the following options: Low-Band (operates in frequencies below 1 GHz), Mid-Band (in the 1–10 GHz range), High-Band (in 20–100 GHz range).

5G will affect our daily communication and mobile usage habits as well as reshape the industry. For instance, due to its video streaming capacity, it will change the way of journalism, etc. Now a single journalist can be deployed on the field with a 5G connected camera to do interviews instead of a news crew with heavy equipment requirements. Moreover, 5G will contribute to adopting advanced technologies such as self-driving cars and smart things (home, factory, city, etc.).

Figure 1.10 illustrates an architectural view of 5G communication. The figure divides the network into two parts, the user data part (also known as the user

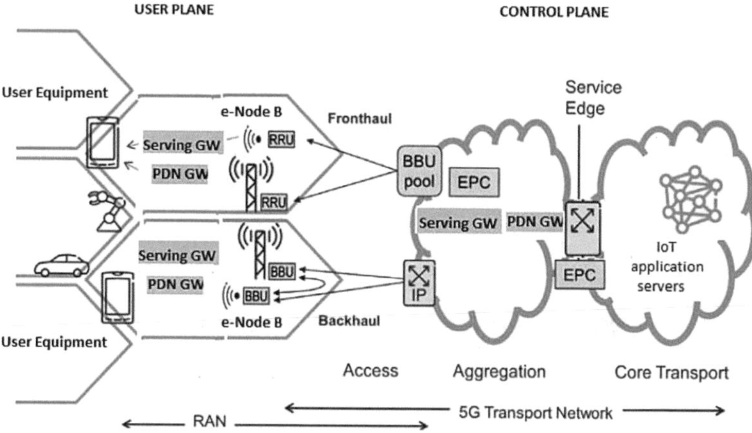

Fig. 1.10 5G reference architecture

plane) and the signaling part (also known as the control plane). This separates their concerns as well as makes the scaling independent. The control plane is supported by the 5G transport network and the latter by the Radio Access Network (RAN). Additionally, the former is further divided into fronthaul and backhaul packet networks. Backhaul is the linkage between a base station and the core wired network, and is often fiber or coax, and in some cases broadband, proprietary wireless links. In most cases the backhaul network is supported by wired communications to enable less communication latency. The front hall network provides the connection between the cell tower radio itself (Radio Head or RH) and the mobile network control backbone (the Baseband Unit or BBU) and CPRI is a well-known standard for this interconnection. The control plane includes additionally the communication with the IoT application servers and the EPC. Furthermore, the EPC comprises by two gateways the Serving and the Packet Data Network (PDN) gateway, serving as the Control Plane for the network. The former is responsible for routing the incoming and outgoing IP packets and the latter serves as a connection point between the EPC and the external IP networks, called as PDN. The PDN gateway routes packets to and from the PDNs and performs various functions such as IP address/IP prefix allocation or policy control and charging. The Serving gateway is logically connected to the PDN gateway and even though 3GPP describes them independently, in practice they may be combined in a single hardware by network vendors. Packets in 5G communication are exchanged between the cloud (application servers of Fig. 1.10) and the control plane. The control plane is afterwards using the standard TCP, UDP and IP protocols to exchange packets with the user plane. The BBU is used to form the evolved (Evolved Node B (e-NodeB) in Fig. 1.10) the main point responsible for the transmission/reception of IP packets to/from the control plane. The BBU also performs packet demodulation as well as amplification to transmit them to the User Equipment (UE), which denotes the end devices used for communication. Connectivity between the user equipment UE and the core network is provided by the E-UTRAN. The E-UTRAN is a collective term for the network and equipment that connects mobile handsets to the public telephone network or the Internet. The user plane contains the e-NodeB and UE consists of three sub-layers: Packet Data Convergence Protocol (PDCP), Radio Link Control (RLC) and Medium Access Control (MAC). In the same figure we can also observe data exchange through UDP between the UE and the Serving gateway, which serves as a relay for the EPC.

5G specifications and protocol details are explained further in Chaps. 2 (2.2.3) and 4 (4.3.4).

1.3.8 Cloud, Edge, and Fog Computing

In *Cloud Computing*, collected data is streamed to a central server and data is processed on a powerful central server which is usually far away from the data source. Cloud technology provides various services like IaaS (Infrastructure as a

Service), PaaS (Software as a Service) and SaaS (Software as a Service). Usually, cloud computing provides highly scalable and almost unlimited storage for business solutions. It also provides really high processing power when compared to the Fog/Edge computing. High latency, possible downtime, and security could be chronic issues in cloud computing for IoT.

Edge Computing brings processing power closer to where the data is being generated. It does not send generated data to a central location directly. The data can be processed on the device where the sensors are connected or a gateway device in close proximity. *Fog Computing* is very similar to the edge computing, as it can be considered as an extension which sets standards for edge computing and specifies how edge computing should work.

Fog/Edge Computing and storage systems are located closer to the edge to eliminate latency. Fog/Edge computing can be integrated with cloud computing if needed. They can work as an extension of cloud computing. The concept of Fog/Edge is harmoniously compatible with IoT. The main goal here is to reduce the amount of sent data to reduce latency between the nodes to improve the overall system response time. Since the data is not transferred to a central location, it is considered more secure. Autonomous vehicles, like self-driving cars, need instant decision to avoid collusion or detecting pedestrians so the sensor or camera data needs to be analyzed on the vehicle (edge) to eliminate the latency or preserve availability. If there is a data link between the vehicle and a remote location, there is always a possibility of a breach or hijacking of the vehicle. It is known[11] that the Iranian cyberwarfare unit once successfully hijacked interrupting the data-link on RQ-170 Sentinel UAV manufactured by Lockheed Martin on December 5, 2011.

In avionics, many sensors are embedded on the aircrafts such as pressure, speed, engine, position, load, fuel, position, torque, steering, engine, gravity, etc. Considering commercial airlines to haul passengers we can say that data needs to be collected from airplane sensors and needs to converted from analog to digital to inform the pilots for decisions.[12] This example shows how the Fog/Edge paradigm is used in commercial airlines since the data is processed on aircraft. However, considering Unmanned Aerial Vehicles (UAV) could be a good fit in cloud computing since UAV is operated from a central location based on the streamed sensor data from UAV sensors. Another example of cloud computing could be from space crafts and orbital satellites. Due to the hardware constraints, collected data needs to be streamed to earth for further processing and analysis. A detailed comparison between cloud and edge/fog computing shown in Table 1.2.

[11]More information available at: https://en.wikipedia.org/wiki/Iran%E2%80%93U.S._RQ-170_incident.

[12]More information available at: https://www.azosensors.com/article.aspx?ArticleID=1614.

Table 1.2 Comparison of cloud, fog and edge computing concepts [13]

Feature	Cloud computing	Fog/Edge computing
Access	Wired or wireless	Wireless
Access to the service	Through server	At the edge device
Availability	Mostly available	Mostly volatile
Bandwidth usage	High	Low
Capacity—Computing	Higher	Lower
Capacity—Storage	Higher	Lower
Connectivity	Internet	Many protocols (Fig 1.5)
Content distributed to	Edge device	Anywhere
Content generator	Man made	Sensor made
Content generation at	Central server	Edge device
Control	Centralized	Distributed
Data analysis	Long term	Instant/Short term
Data processing	Far from data source	Closer to data source
Latency	High	Minor
Location of resources (i.e. processing and storage)	Center	Edge
Scalability	High	Low
Security	Weaker	Stronger
Mobility	Limited	Supported
Number of users	Millions	Billions
Virtual infrastructure location	Enterprise server	Enterprise/User devices

1.3.9 Data Streaming

Data streaming is related with time series and optimal for detecting patterns over time. For instance, tracking and recording the length of a web session sent for a client side. Most of the IoT data is also well-suited to data streaming paradigm. Applications like security cameras, traffic sensors, crowd sensors, health sensors, activity logs, and transaction logs are all good examples for data streaming.

Data streaming allows analyzing data in real time and gives insights into a wide range of activities, such as metering, server activity, geolocation of devices, or website clicks; by using some advanced computing such as real-time aggregation and correlation, filtering, or sampling. For example, in a smart grid network scenario, the AMI can monitor instant and cumulative throughput of the users and generates alerts when certain thresholds are reached [1].

As discussed earlier, with the vast amount of data (TBs) being created by the collection of AMI meters in a smart grid system, bottlenecks will occur in processing and storing this enormous big data. The solution to this problem might be processing the data closer to the source (smart meters) as much as possible. Data streaming algorithms employed at the edge of the network can help in this manner.

1.3.9.1 Data Handling and Computing in Information Systems

This sub-section overviews data handling and computing. There are two types of data handling for information systems in general [41]:

- **One-time query:** Traditional Data-Base Management Systems (DBMSs) have been used for decades to manage data. The primary goal of DBMSs is to store data in a form of persistent data-set and then run one-time queries over it.
- **Continuous query:** Data stream processing is one pass analysis over the data on the fly. In contrast to DBMSs, stream processing employs continuous queries which are queries that are issued once and continuously run over the flow of tuples (new data records).

As shown in the Fig. 1.11, stream processing can run the query immediately after a new tuple arrives which in turn enables online analysis. Although DBMS is useful in applications where the updates of the database are relatively infrequent, it is inefficient for modern high-volume (data size)/high-velocity (the speed of data generation or arrival)/high-veracity (discrepancy)/high-variety (various data structures) data-driven applications where Big Data (4 V's) is being generated. At that point, continuous query will be an efficient solution to be used for Big Data generating systems.

Authors of this chapter projects that, the traditional technique of *one-time query* will be replaced by the *continuous query* technique in the near future for the industrial networks, as it will be more beneficial not only for the network provider but also for the industrial users.

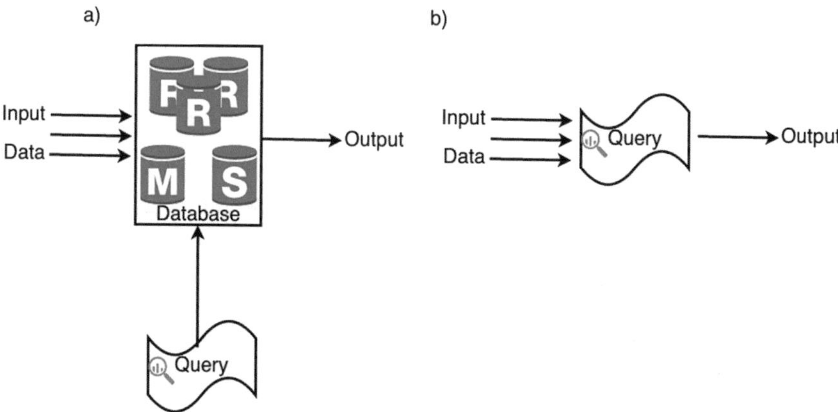

Fig. 1.11 Data handling in information systems. Here, *R* represents replica or reporting database (DB) instances, *S* represents standby DB, and *M* represents the active DB. (**a**) One time query DBMS (**b**) Continuous query in stream processing

1.3.9.2 How the Data Is Processed?

- **Data fusion:** Data fusion is the process of integrating data destined from multiple sources to produce more consistent, accurate, and useful information.
- **Sensor fusion:** A specific case of data fusion, in which data resourced from many sensors are incorporated into a meaningful and useful form.
- **Data streaming:** Continuously generated data by different sources should be processed incrementally using *data stream processing techniques* without having access to all of the data being produced or transmitted. It is usually used in the context of big data in which it is generated by various sources at high speed.

1.3.9.3 Data Streaming Tools

As the data streaming concept grows exponentially, a number of tools have already gained popularity among implementer and coders [1]:

Amazon Kinesis Firehose: It is a managed, scalable, cloud-based service which allows real-time processing of large data streams.

Apache Kafka: It is a distributed publish-subscribe messaging system which integrates applications and data streams all-together.

Apache Flink: It is a Stream Processing Engine (SPE) which works on streaming data flow and provides facilities for distributed computation over data streams.

Apache Storm: It is a distributed real-time computation system. Storm is used for distributed machine learning, real-time analytics, and numerous other cases, specifically with high data velocity and veracity.

1.3.10 Digital Twin

As discussed in [33], IIoT leads to the Industry 4.0, by improving the factory environments by enhancing the user interactions in between the machines and operators. For instance, *Digital Twin* is one of the most intriguing technological developments that will shape tomorrow's industrial environments.

Especially, as foreseen in [49], Product Life-cycle Management (PLM) of industrial environments will benefit from *Digital Twin* concept. It will not only ease the way of overall production monitoring, but also enhance the predictability of the events of malfunctions, bottlenecks in the production line, etc. Up-to date, PLM is very time-consuming and challenging in terms of manufacturing, sustainability, efficiency, intelligence, and service phases of the product design. The digital twin will enable manufacturing operators to have a digital replica of all of their production lines through the entire product life-cycle. Hundreds of sensors will be placed in the manufacturing process environment, to collect data from various dimensions such as ambient conditions, operational characteristics of the machines along with their working efficiency on the job they are performing. All this data is continuously communicating and collected by the digital twin [51].

In [32], it is clearly mentioned that: "As the technology trend of industrial IoT increases, digital twin technology is more significant now than ever before." Owing to the improvements in IoT and especially on IIoT, digital twin became very affordable and has an enormous potential to realize the future of the manufacturing industry. The benefit provided to engineers is the fact that real-world products can be designed and tested virtually by the digital twin. Product maintenance and management can be drastically improved if the real 'thing' of the production environment to be replaced by the digital twin within real-time precision.

IIoT is en enabling technology for Digital Twin by providing the functionalities of remote sensing of the sensors and remote controlling of the actuators. As shown in Fig. 1.12, IIoT provides the means of communications to provide the link in between the real factory machinery and the digital twin equivalent at the command center of the factory.

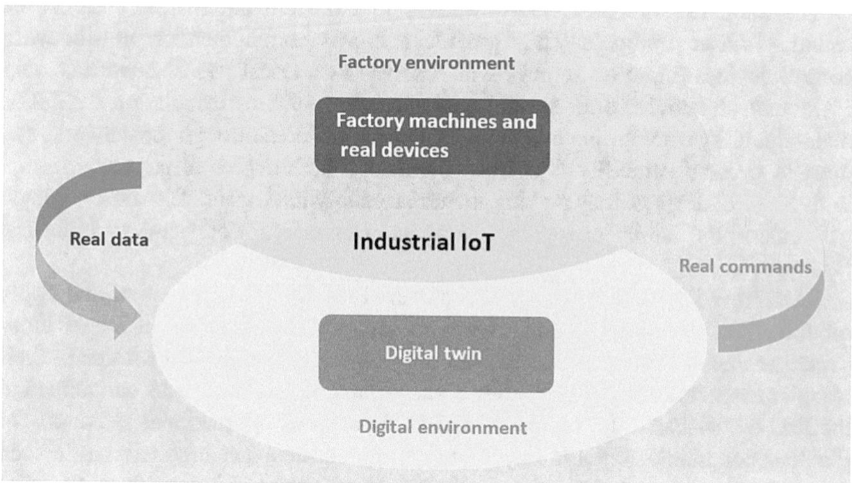

Fig. 1.12 IIoT and Digital Twin concept for tomorrow's factory environments

1.4 Enabling Technologies for Industrial Networks

This section[13] thoroughly describes most (if not all) of the enabling technologies for industrial and IIoT networks, namely as follows: Ethernet, IP, PROFINET, CBA, Modbus, TCP, CANopen, NMT, PDO, SDO, Ethernet Powerlink, EtherCAT, EtherCAT, V2V, V2X, V2I,I2V, V2P, P2V, V2N, N2V, I2N, N2I, DLMS/COSEM, DNP3, ZigBee, and BACnet.

1.4.1 Recent Developments in Industrial Networks for Industry 4.0

Industrial automation systems are used in manufacturing, quality control and material handling processes. General purpose controllers for industrial processes include Programmable Logic Controller (PLC) devices as well as sensors and actuators. The exchanged data in industrial processes are stored in powerful computers, such as servers. The main concern in such systems is to assure real-time performance as well as efficiency in terms of resource usage, such as the energy or memory consumption. Typical examples of such systems are distributed control systems or safety critical systems.

The main technologies used nowadays in industrial automation systems are called *Fieldbus protocols*. They provide a digital communication link between control devices (input or output), which serves as a Local Area Network (LAN). Fieldbus technologies offer several characteristics, such as installation flexibility, maintainability (monitoring and maintenance are handled through the network) and most of all configurability. The latter provides a high degree of parameterization in the control devices, thus making them reasonably intelligent. The most common solutions in the family of Fieldbus protocols rely on the Real-Time or Industrial Ethernet [22].

Real-Time Ethernet is using the standard Ethernet communication and apply modifications to extend it with real-time capabilities. Currently, a lot of Real-Time Ethernet solutions are in use, but only some of them are known due to their technical aspects and standardization status. Many of these solutions are defined in the IEC 61784-Part 1 [47] and IEC 61784-Part 2 [48] international standards for Fieldbus communication and rely mainly on the master/slave architecture. In such an architecture a particular device manages the network and has uniformal control over the other devices.

The Real-Time Ethernet that employ a master/slave architecture are classified into three categories according to the implementation of the slave devices in the

[13]This section is partially extracted from the unpublished part of the Ph.D. thesis of Dr. Lekidis [36], the full version of which is available at: https://tel.archives-ouvertes.fr/tel-01261936v2/document.

system. We hereby present these categories by evenly giving characteristic examples of technologies that are mainly described by the IEC 61784-Part I and IEC 61784-Part II international standards for each one of them. Moreover, for solutions that are not included in these standards, supporting material is provided.

The first category is using the TCP/IP protocol stack and hardware, such as the standard Ethernet controller as well as Ethernet switches. However, it does not provide guarantees for real-time performance as the communication latencies deriving from the use of switches as well as of the best-effort delivery service are unpredictable and result in an average data rate of 100 ms. Typical technology variants belonging in this category include Ethernet/IP, PROFINET Component Based Automation (CBA) and Modbus/TCP. The second category uses the same hardware, but employs an additional timing layer in the third layer (Internet) of the TCP/IP stack, in order to control access to the medium. Technology variants belonging in this category include PROFINET Real-Time (RT) and Ethernet POWERLINK (EPL) (see Sect. 1.4.2.8). An important feature of this category is that it provides better real-time performance (average data rate below 10 ms), which can be additionally ameliorated as some of the related technologies are also deployed using Ethernet hubs (e.g. Ethernet POWERLINK). Finally, the third category aims on achieving the best possible real-time performance for the most demanding class of applications. Nevertheless, this is not feasible without specific modifications on the underlying hardware. These modifications depend on the technology and can either concern the Ethernet controller or the Ethernet switches. Technologies related to this category include PROFINET Isochronous Real Time (IRT), SERCOS III, EtherCAT and TTEthernet [50]. The selection of the category as well as the specific master/slave solution for an application depends on its requirements and needs.

Even though Real-Time Ethernet technologies are widely used for industrial automation systems, application development is still challenging, due to their low level complexity as well as their high expertise needed for their configuration. Therefore, a higher layer of abstraction is required, which is typically found in application-layer protocols. An increasingly popular application-layer Fieldbus protocol is CANopen (see Sect. 1.4.2.1), as it provides a vast variety of communication mechanisms, such as time or event-driven, synchronous or asynchronous as well as additional support for time synchronization and network management. Furthermore, it offers a high-degree of configuration flexibility, requires limited resources and has therefore been deployed on many existing embedded devices.

1.4.2 Industrial Automation Protocols

1.4.2.1 CANopen

CANopen [18] is an increasingly popular application layer protocol, belonging to the family of Fieldbus protocols for networked embedded systems. Its main attributes are the vast variety of communication mechanisms, such as time or

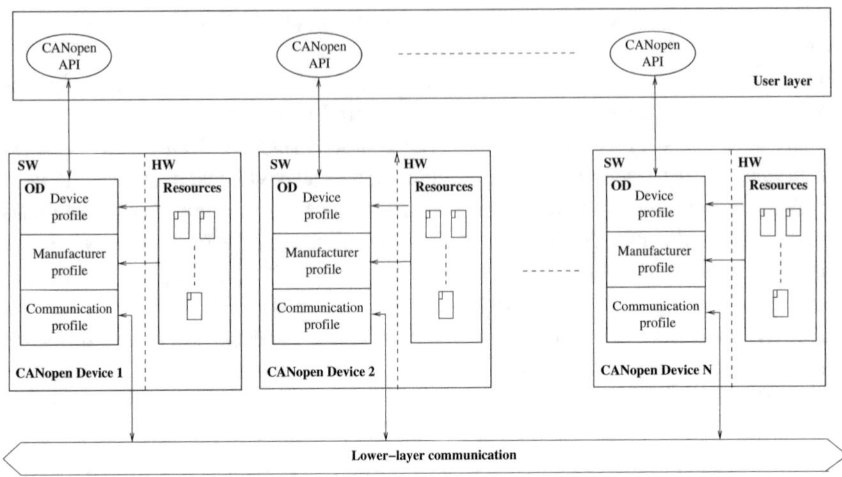

Fig. 1.13 Communication in a CANopen system

event-driven, synchronous or asynchronous as well as the support for time synchronization and network management mechanisms. Additionally, it provides a high-degree of configuration flexibility and requires limited resources. CANopen uses a master/slave architecture for management services, but concurrently allows the utilization of the client/server communication model for configuration services as well as the the producer/consumer model for real-time communication services. A comprehensive introduction to the protocol can be found in [42]. Unlike other Fieldbus protocols it does not require a single master controlling all the network communication. Instead a CANopen system is specified by a set of devices (Fig. 1.13), which in turn use a set of profiles, in order to define the device-specific functionality along with all the supported communication mechanisms. The communication profile defines all the services that can be used for communication and the device profile how the device-specific functionality is made accessible. The communication profile is defined in the DS-301 standard [18], whereas the device profiles providing a detailed description on CANopen's usage for a particular application-domain, are defined in the DS-4xx standards.[14] If CANopen systems require configurations or data access mechanisms not covered by the standard communication profile, profile extensions can also be defined. These are called Frameworks and are found in the DS-3xx standards (see footnote 15).

The protocol's communication mechanisms according to the DS-301 are specified by standard *Communication Objects (COB)*. All the COBs have their own priority and are transmitted through regular frames of the chosen lower-layer protocol. They are generally divided in the following main categories:

[14]http://www.can-cia.org/index.php?id=440.

Table 1.3 Frame fields in the CAN HW/Communication model

Index	Data type
1	BOOLEAN
2	INTEGER8
3	INTEGER16
4	INTEGER32
5	UNSIGNED8
6	UNSIGNED16
7	UNSIGNED32

- *Network Management objects (NMT)*, used for the initialization, configuration and supervision of the network
- *Process Data Object (PDO)*, used for real-time critical data exchange
- *Service Data Object (SDO)*, used for service/configuration data exchange
- *Predefined objects*, specifying standard object that are included in every device. The featured objects in this category are:
  - *Synchronization object (SYNC)*, broadcasted periodically to offer synchronized communication as well as coordinate operations
  - *Timestamp object (TIME)*, broadcasted asynchronously to provide accurate clock synchronization using a common time reference
  - *Emergency object (EMCY)*, triggering interrupt-type notifications whenever device errors are detected

All the aforementioned objects are stored in a centralized repository, called Object Dictionary (OD), which holds all network-accessible data and is unique for every device. Commonly used to describe the behavior of a device, it supports up to 65,536 objects addressed through a 16-bit index. The COBs are spread to distinct areas, defining communication, manufacturer and device specific parameters (Fig. 1.13). The latter are left empty and are used by manufacturers, in order to provide their own device functionalities. Each OD entries has also an associated data type with a respective code, used to identify it. The supported data types along with their code are given by the following table (Table 1.3).

Furthermore, every CANopen object in the OD has a dedicated type with respect to the information it stores. and each type has an associated code. In particular, the CANopen object may be either a variable (object code 7), an array (object code 8) or a record (object code 9). The difference between the array and the record is that the former contains sub-indexes of the same data type, whereas the latter of different data types.

The OD entries are described by electronically readable file formats, such that they are uniformly interpreted by configuration tools and monitors. According to the DS-306 standard [16] they are provided by the INI format files and termed as Electronic Data Sheet (EDS) files. These files provide a generic description of a device type. However, since CANopen allows parametrization according to manufacturer specifications, a specific file format exists and is defined as Device

Configuration File (DCF). This file describes the configuration for a specific device. Nevertheless, EDS and DCF files have limitations on the validation and presentation of the data as well as require a specific editor. Therefore, new XML-based device descriptions were introduced according to the DS-311 standard [17]. These substitute the EDS with the XML Device Description (XDD) file format and the DCF with the XML Device Configuration (XDC) file format. A fragment of an XDC a device description is provided in Example 1.1. Currently, the protocol supports both device descriptions.

We hereby describe thoroughly the CANopen objects, according to the classification we have previously mentioned.

1.4.2.2 Network Management (NMT) Objects

The NMT objects are generally transmitted by devices, which act as an NMT master in CANopen. Upon the reception of such object a CANopen device is informed to transit in a different NMT state. Each NMT state supports specific communication mechanisms and objects of the CANopen protocol, related to the device functionality. The device can switch between three main state, named Pre-Operational, Operational and Stopped. In the Pre-Operational state a device can actively participate in all communication mechanisms related to SDO and Predefined object exchange. However, the main difference with the Operational state is that it doesn't support PDO object exchange. In the Operational state the device is fully operational and can perform all the functionalities that it was designed to do. The NMT master can also switch off the device by transmitting a dedicated NMT object. Accordingly, the device has to switch to the Stopped state stopping all communication, except from the support for the reception of NMT objects.

1.4.2.3 Process Data Objects (PDO)

The real-time data-oriented communication follows the producer/consumer model. It is used for the transmission of small amount of time critical data. PDOs can transfer up to 8 bytes (64 bits) of data per frame and are divided in two types: The transmit PDO (TPDO) denoting data transmission and the receive PDO (RPDO) denoting data reception. Therefore, a TPDO transmitted from a CANopen device is received as an RPDO in another device (Fig. 1.14). Additionally, the supported scheduling modes are:

- *Event driven*, where the transmission is asynchronous and triggered by the occurrence of an object-specific event
- *Time driven*, where transmission is triggered periodically by an elapsed timer
- *Synchronous transmission*, triggered by the reception of the SYNC object, further divided in:

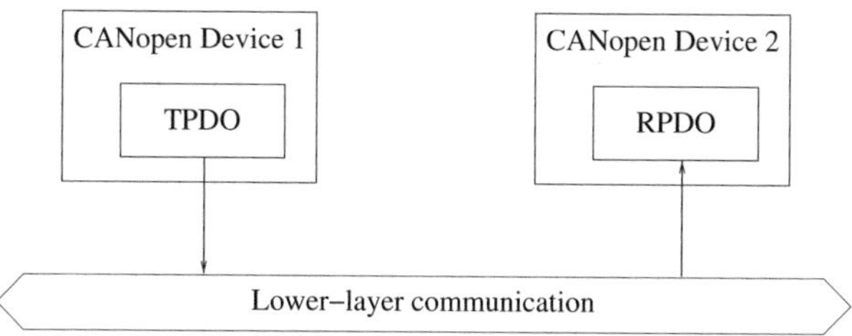

Fig. 1.14 PDO communication

  - Periodic transmission within an OD-defined window (synchronous window), termed as *Cyclic PDO* transmission
  - Aperiodic transmission according to an application specific event, termed as *Acyclic PDO* transmission

- *Individual polling*, triggered by the reception of a remote request (see [15])

Each PDO is described by two OD sub-objects: The Communication Parameter and Mapping Parameter. For a TPDO (e.g. OD entry 1800h and 1A00h respectively) the former indicates the way it is transmitted in the network and the latter the location of the OD entry/entries, which are mapped in the payload. On the contrary for a RPDO (e.g. OD entry 1400h and 1600h respectively) the former indicates how it is received from the network and the latter the decoding of the received payload and the OD entry/entries where the data is stored.

The Communication Parameter entry includes the *Communication Identifier (COB-ID)* of the specific PDO, the scheduling method, termed as *transmission type*, the *inhibit time* and the *event timer*. The inhibit time (expressed as a multiple of 100 μs) defines the shortest and the event timer (expressed as a multiple of 1 ms) the longest time duration between two consecutive transmissions of the same PDO.

The Mapping Parameter describes the structure of a PDO. It can be of two types, that is, static or dynamic. Static mapping in a device cannot be changed, whereas dynamic mapping can be configured at all times through an SDO.

In Table 1.4 we illustrate the sample configuration and mapping parameters of a TPDO for a CANopen device. They represent how analogue input data, obtained from temperature sensors, are described in the OD of the device. Accordingly, we also present the fragment, which describes them in an XDC format.

Table 1.4 Example TPDO configuration and mapping parameters in the OD

Index	Subindex	Description	Value
1800h (6144)	0	Number of entries	5
	1	COB-ID	641 + deviceID
	2	Transmission type	255
	3	Inhibit time (in ms)	1
	4	Reserved	–
	5	Event timer (in ms)	1000
1A00h (6656)	0	Number of entries	1
	1	1st object to be mapped	6400h (25,600)/Subindex 1
6400h (25,600)	0	Number of analogue inputs	n
	1	input 1 (in °C)	30.5
::::	::::	::::	::::
	n	input n (in °C)	23

Example 1.1 The fragment of "Program Code" provided in the next page, illustrates the TPDO configuration of Table 1.4 in an XDC format.

Program Code

TPDO configuration in an XDC CANopen specification:

```
<ProfileBody>
<ApplicationLayers>
<CANopenObjectList>
<CANopenObject index="1800" name="1st Transmit PDO Communication Parameter"
   objectType="9" subNumber="5">
<CANopenSubObject subIndex="00" name="Number of entries" objectType="7"
   dataType="0005" lowLimit="0x02" highLimit="0x05" accessType="ro"
   defaultValue="5" PDOmapping="no" actualValue="5"/>
<CANopenSubObject subIndex="01" name="COB-ID" objectType="7"
   dataType="0007" PDOmapping="no" uniqueIDRef="UID_PARAM_180001"/>
<CANopenSubObject subIndex="02" name="Transmission Type" objectType="7"
   dataType="0005" PDOmapping="no" uniqueIDRef="UID_PARAM_180002"/>
<CANopenSubObject subIndex="03" name="Inhibit Time" objectType="7"
   dataType="0006" PDOmapping="no" uniqueIDRef="UID_PARAM_180003"/>
<CANopenSubObject subIndex="05" name="Event Timer" objectType="7"
   dataType="0006" PDOmapping="no" uniqueIDRef="UID_PARAM_180005"/>
.....
<CANopenObject>
<CANopenObject index="1a00" name="1st Transmit PDO Mapping Parameter"
   objectType="9" subNumber="9">
<CANopenSubObject subIndex="00" name="Number of entries" objectType="7"
   dataType="0005" accessType="ro" defaultValue="1" PDOmapping="no"
   actualValue="8"/>
<CANopenSubObject subIndex="01" name="PDO Mapping Entry" objectType="7"
   dataType="0007" PDOmapping="no" uniqueIDRef="UID_PARAM_1a0001"/>
</CANopenObject>
<CANopenObject index="6400" name="Read Analog Input 16-bit"
   objectType="8" subNumber="13">
<CANopenSubObject subIndex="00" name="Number of elements" objectType="7"
   dataType="0005" accessType="ro" defaultValue="12" PDOmapping="no"
   actualValue="N"/>
<CANopenSubObject subIndex="01" name="AnalogInput16_1" objectType="7"
   dataType="0003" PDOmapping="TPDO" uniqueIDRef="UID_PARAM_640101"/>
.....
```

```
<CANopenSubObject subIndex="01" name="AnalogInput16_N" objectType="7"
   dataType="0003" PDOmapping="TPDO" uniqueIDRef="UID_PARAM_64010N"/>
</ApplicationLayers>
.....
</ProfileBody>
```

We can observe that all the CANopen objects are defined inside the construct "CANopenObjectList". Moreover, for each object ("CANopenObject") dedicated elements are also defined to denote the sub-objects ("CANopenSubObject"). Every sub-object has an index ("subIndex"), a name ("name"), a specific object code ("objectType") and a type of data that it may contain ("dataType"). Additionally, a sub-object includes a default value ("defaultValue") and its specific value in the application ("actualValue"). Nevertheless, the latter is optional and if it not provided in the XDC file, it is accordingly set equal to the default value.

1.4.2.4 Service Data Objects (SDO)

The service oriented communication follows the client/server model. It supports large, non-critical data transfers and uses three modes to allow peer-to-peer asynchronous communication through the use of virtual channels:

- *Expedited transfer*, where service data up to 4 bytes are transmitted in a single request/response pair.
- *Segmented transfer*, where service data are transmitted in a variable number of request/response pairs, termed as segments. In particular it consists of an initiation request/response followed by 8-byte request-response segments.
- *Block transfer*, optionally used for the transmission of large amounts of data as a sequence of blocks, where each one contains up to 127 segments.

A CANopen device can either receive or request an SDO, therefore these objects are not separated as the PDOs, instead they are distinguished according to their identifiers (Sect. 1.4.2.6). The communication is always initiated by a device defined as client in the network towards the server, nonetheless information is exchanged bidirectionally with two services: *Download* and *Upload*. The former is used when the client is attempting service data transmission to the server, whereas the latter when it is requesting data from the server. In both services the use of the virtual channel ensures that the received SDO is identical to the transmitted (Fig. 1.15), unlike in PDO communication. Each request or receive SDO uses byte 0 as metadata, containing important information about the transmitted object and reducing the payload to seven bytes per frame. This byte includes the command specifier, which indicates the type of the frame, that is initiate or domain segment and request or response. The command specifier is either termed as client command specifier (*ccs*) for the client device or server command specifier (*scs*) for the server device. For the initial request/response pair byte 0 also determines which of the three

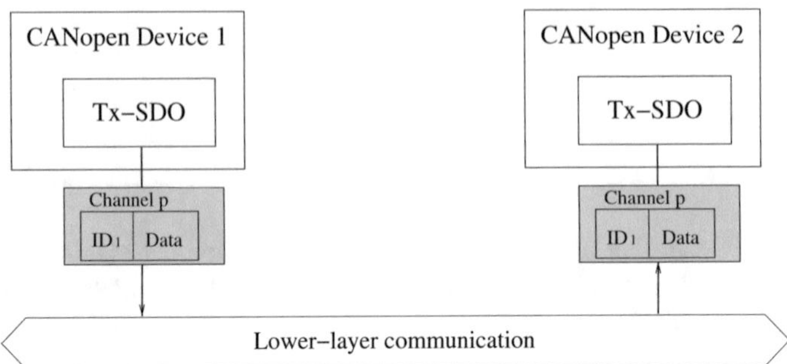

Fig. 1.15 SDO communication

modes is used (see [18]). If transmission errors are detected either on the client or the server side, data transfer is aborted through the SDO abort frame. SDOs are used for configuration and parametrization, but also allow the transmission of a large quantity of asynchronous data, consequently they are always assigned a lower priority than PDOs.

1.4.2.5 Predefined Objects

These specific objects provide additional functionalities to the protocol. Their transmission is following the producer/consumer communication model. Particularly, the SYNC and the TIME object are always transmitted from a specific device (Producer), according to the OD specification, whereas the EMCY object can be transmitted by any device in the network (dynamical configuration). The Predefined objects are always assigned with a high priority, in order to be transmitted as soon as possible.

The SYNC object is used to enable synchronized operation. Yet, if the transmission is handled by the CAN protocol the derived delays due to non-preemption can result to a certain jitter. Thus, if it does not provide the required accuracy for the synchronization, CANopen enables the use of the TIME object, containing a reference clock time. Though implementing a different synchronization mechanism, this object is used for accuracy, measuring the difference between theoretical and the actual transmission time of the SYNC and transmit it through a subsequent PDO.

The EMCY object is used in internal error conditions in a device and transmitted as an interrupt, in order to notify other devices. However, no notification is present when the internal error is fixed and thus the other devices cannot know the change of condition. Consequently, its implementation is not considered mandatory in CANopen systems.

Table 1.5 Predefined connection set

Communication object	COB-ID
NMT	0
SYNC	128
EMCY	129
TIME	256
TPDO1	385–511
RPDO1	513–639
TPDO2	641–767
RPDO2	769–895
TPDO3	897–1023
RPDO3	1025–1151
TPDO4	1153–1279
RPDO4	1281–1407
Tx-SDO	1408–1535
Rx-SDO	1536–1663

1.4.2.6 Network Configuration

Considerable complexity in CANopen systems is found in the configuration and allocation of a frame identifier to each COB. As CANopen allows parametrization the allocation scheme can be configured according to specific manufacturer requirements, however sufficient attention must be given to the priority group of each object. Therefore, to reduce the complexity of CANopen system development, a default allocation scheme is provided for applications using CAN as the lower-layer communication protocol. This scheme is named Predefined Connection Set. As defined by this scheme, every object is assigned an identifier (COB-ID) according to Table 1.5, derived from its priority in the protocol. Nevertheless, each frame has its own identifier, since the COB-ID is augmented by the specific identifier of the node transmitting it. Every device can use up to four TPDOs, four RPDOs, one EMCY and one SDO. All the COB-IDs can be configured, except of the SDOs, if the particular device allows it.

1.4.2.7 Application Development with CANopen

Apart from its use in automotive systems as a high-level protocol on top of CAN, CANopen can be used as an application layer protocol in industrial automation systems where it is integrated with Real-Time Ethernet technologies. The integration with many of those technologies is facilitated by the existence of a gateway [53] or a proxy [2]. Nevertheless, a possible constraint in this case are the additional latencies that often have a strong impact in the real-time performance. Therefore, a direct integration of CANopen as an application layer protocol along with its communication and device profiles is often preferred as an industrial solution. Two technologies facilitating this integration are Ethernet POWERLINK

(EPL) (Sect. 1.4.2.8) and EtherCAT [43]. In particular, it can be integrated with many of Real-Time Ethernet technologies, however this usually implies the. Both technologies support fully the CANopen communication profile, in order to describe its services and mechanisms into a real-time Ethernet environment. Nevertheless, in many implementations EPL is preferred over EtherCAT, since it does not require any specific hardware modifications and is more suitable for the transmission of large amounts of data.

1.4.2.8 Ethernet Powerlink (EPL)

ETHERNET Powerlink (EPL) [48] is a commercial protocol for industrial automation systems based on the Fast Ethernet IEEE 802.3. One of protocol's major advantages is that it can operate with either the use of Ethernet switches or hubs, depending on the temporal constraints of the application. To overcome the effect of collisions occurring in standard Ethernet systems, EPL uses a TDMA technique (deployed in the data link layer), which is based on a mixed polling and time slicing mechanism, called Slot Communication Network Management (SCNM) (Fig. 1.16). This technique uses a special node, referred as Managing Node (MN), to grant the slave devices, referred as Controlled Nodes (CN's), access to the medium only when they are polled. The use of SCNM hampers the direct deployment of standard Ethernet devices in the network, as they would corrupt the access mechanism. To overcome this limitation dedicated gateway are connected to control the communication traffic of standard Ethernet devices. The supported topologies in EPL are the line and star topology.

EPL supports periodic and event-based data exchange during a cyclic period of fixed duration. This period is divided in four phases, namely the starting, the isochronous, the asynchronous and the idle phase. The synchronized transition between phases is done through broadcast frames initiated by the MN device. More specifically, the reception of the Start of Cycle (*SoC*) frame by the slave devices ends the starting phase and accordingly begins the isochronous (cyclic) phase. During this phase the MN polls progressively every CN through a *PReq* unicast frame, in order to receive their data responses through the subsequent *PRes* frames. The *PRes* frames are also broadcasted, in order to facilitate data distribution amongst all the

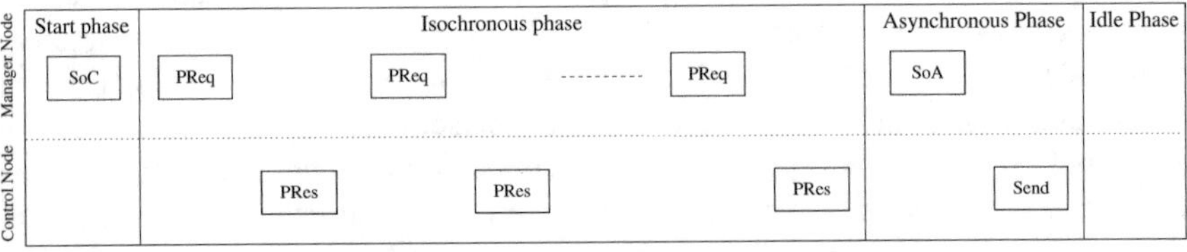

Fig. 1.16 EPL cycle

remaining nodes. Having polled all the CN devices in the EPL network, the MN broadcasts the Start of Asynchronous (*SoA*) frame, to indicate the beginning of the asynchronous period. This period allows a single asynchronous transaction (*Send* in Fig. 1.16) to be performed. This transaction might be an asynchronous EPL data frame (*ASnd* frame), detection of active stations (*IdentRequest* frame), or even a standard Ethernet data frame. All the asynchronous transactions are enqueued in the MN, in order to be transmitted according to their priority. As the asynchronous period is used for the exchange of large frames, the EPL cycle includes the idle phase to ensure that the ongoing transaction has ended.

1.4.2.9 EPL Frame Format

EPL frames (Fig. 1.17) are encapsulated and transmitted in the Data field of IEEE 802.3 standard Ethernet frames. Therefore, their main difference with the legacy Ethernet frames is the Ethernet Type field of the Ethernet frame, which is set to the hexadecimal value 88ABh. An EPL frame consists of five fields: the Message Type, specifying the type of EPL frame (as defined above), the EPL source and destination addresses, the EPL data to be exchanged and an optional padding. The Message Type field can contain one of the values present in Table 1.6.

The EPL source and destination addresses for each frame are specified according to Table 1.7. Specifically, we can denote that for each EPL node there is a unique address, except from the MN which is always equal to 240. Likewise, the CN's have an id in the range 1–239 and 253 is used by a node to address itself. Finally, as the minimum Ethernet II frame size is equal to 64 bytes, extra padding data is added to the packet. The size of the data padding in an EPL frame can be up to 43 bytes.

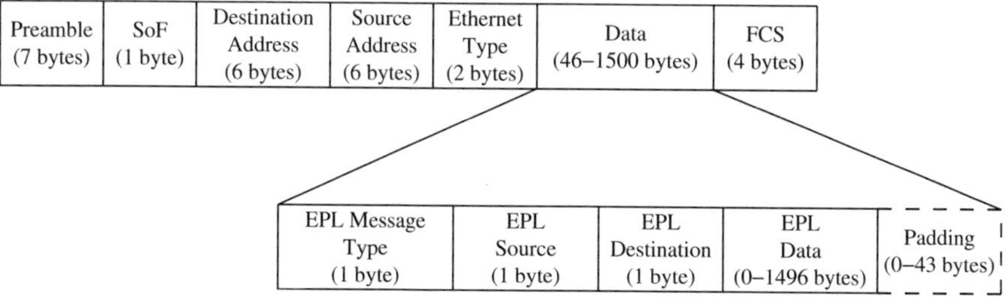

Fig. 1.17 EPL frame format

Table 1.6 Message type field of EPL frame

Message type	EPL frame type
01	Start of Cyclic (SoC)
03	Poll Request (PReq)
04	Poll Response (PRes)
05	Start of Asynchronous (SoA)
06	Send (ASnd or IdentRequest)

Table 1.7 EPL node addressing

Address	EPL node
0	Invalid
1–239	Controlled node (CN)
240	Managing node (MN)
241–252	Reserved
253	Self diagnostic identifier
254	EPL to legacy ethernet router
255	Broadcast identifier

1.4.2.10 Application Layer Profile

EPL is fully integrated with the CANopen protocol as well as its communication and device profiles as presented in Sect. 1.4.2.1. This integration facilitates the description of its services and mechanisms into a real-time Ethernet environment.

As a result of the integration, CANopen's objects are encapsulated into lower-layer EPL frames. Initially, during the isochronous phase of the EPL cycle (Fig. 1.16), data relevant to the application are stored and exchanged through Process Data Objects (PDOs). To this regard, the MN sends a TPDO to each CN via a PReq frame which, in turn, stores the data in one or more RPDOs and responds with a TPDO encapsulated in a PRes frame. Moreover, in the asynchronous phase configuration data are exchanged through Service Data Objects (SDOs), respectively encapsulated in ASnd frames.

EPL also uses the Object Dictionary (OD) to store all the network-accessible data. It may also contain a maximum of 65,536 entries as well distinguished in the communication, manufacturer and device specific categories. The OD entries are described only by the XDD and XDC file formats, similar to the one presented in Example 1.1.

Even though the CANopen communication profile is fully integrated in EPL, there are also minor differences between them as for with the SDO channels, which is are defined and configured during initialization in CANopen, but instead EPL allows a dynamical configuration of these channels. Another difference lies on the transmission of unconfirmed segment frames during the segmented data transfer in EPL, whereas as mentioned in Sect. 1.4.2.1 CANopen segments are always confirmed. A method for confirming the transmission when large amounts of configuration data need to exchanged is through the use of multiple expedited SDO transfers.

1.4.2.11 Application Development with EPL

When developing applications in EPL the most frequent challenges (by priority level) that may arise are:

1. **Separation of functionalities between the EPL nodes.** The developer should be able clarify and implement a different behavior for each EPL node, according to the type of EPL application. As an example in a sense-compute-control application the MN node is not only used for polling, but the CN's may often require dedicated data from it, in order to perform actuations. This is handled in the EPL cycle by supporting transmission capabilities to the MN node using proper configuration. Therefore, the developer should be able clarify and implement a different behavior for each EPL node.
2. **Mapping of application-specific functionality to the Object Dictionary entries.** Once a clear functionality separation is defined, the developer should assigns specific entries to the Object Dictionary for handling the network configuration as well as the exchange of time critical or asynchronous data in the application. This task should be done in respect to the CANopen profile and thus requires high expertise, in order to define the correct data encoding and object linking and may be time consuming if the application's behavior is complex.
3. **Selection of the EPL configuration parameters.** EPL applications are characterized by strict timing constraints. Therefore the selection of parameters, such as the cycle duration, the timeout for acquiring the polling responses, the tolerance timeout in the CN's for receiving the SoC frame and the maximum transmitted data during the asynchronous phase, determines to a large extent the EPL application functionality. The selection of these parameters also depends on the characteristics of resource-constrained devices (e.g. computational platforms), which are chosen in the underlying hardware architecture.

From the aforementioned challenges we can reason that the correct configuration of the MN and the CN devices is of vital importance in EPL application development. To this end, an open source (BSD Licence) tool for the development of applications with Ethernet Powerlink (EPL) was defined, named openPOWERLINK [4]. openPOWERLINK is an open source (BSD Licence) Real-Time Ethernet stack provided by SYSTEC Electronic.[15] openPOWERLINK is developed using a layered approach, which segments the system in a hierarchical way, namely the user and the kernel part. The former implements the application layer of the EPL protocol and provides an API for the development of EPL applications. It contains an implementation for the OD, as well as the PDO, SDO, Error Handler and Event Handling modules. The latter implements the Data Link Layer (DLL) of the EPL protocol and the necessary drivers to communicate with the hardware. It also contains an Event Handling module as well as implementations for an Ethernet and a time-critical driver (for the time slicing mechanism). The Event Handling module is responsible for delivering events, which are related to object dictionary accesses, completion of SDO transfers, configuration and stack errors etc. The two parts interact with each other by message passing through the Communication

[15]http://www.systec-electronic.com.

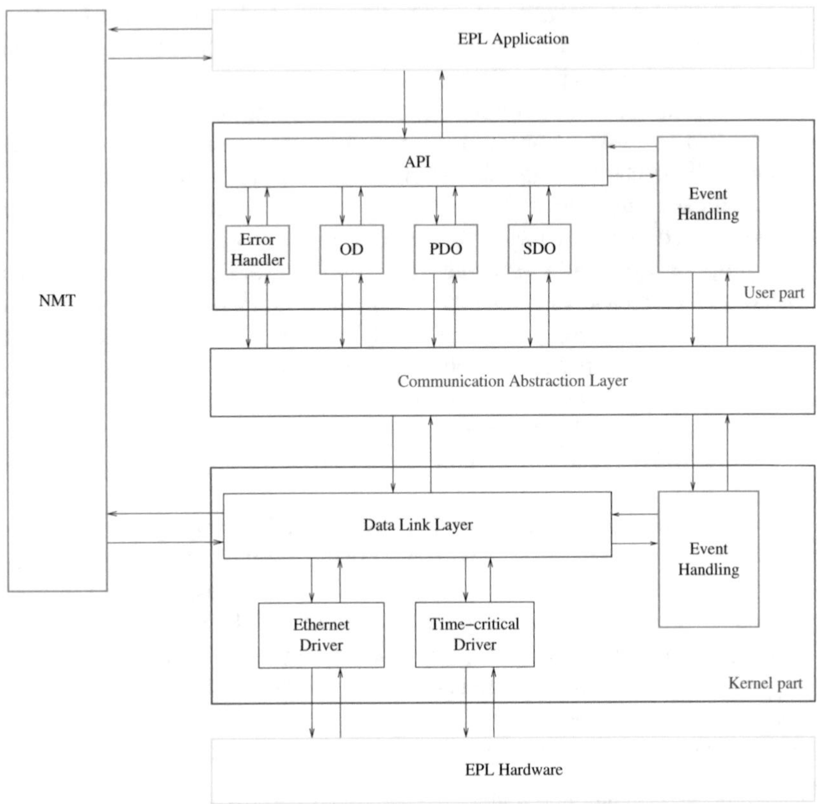

Fig. 1.18 openPOWERLINK stack architecture

Abstraction Layer (CAL). All the processes defined above the CAL have a high-priority in the stack, whereas the ones below have a low-priority. The overall architecture of the openPOWERLINK stack is illustrated in Fig. 1.18. In order to allow NMT functionalities related to the CANopen protocol (Sect. 1.4.2.2) openPOWERLINK supports an additional NMT module to manage the NMT state machine. The Managing Node can use this module to set its or the Controlled Nodes' state machines into four states, namely PreOperational1, PreOperational2, ReadyToOperate and Operational. In the PreOperational1 state all the modules are stopped, the PreOperational2 allows the functionality all the modules except from the PDO and the ReadyToOperate is a transitional state where the PDO module before moving to the Operational state.

openPOWERLINK handles communication between different layers of the stack using dedicated methods, such as *variable linking*. This method defines specific API variables, called *process variables*, which are accordingly linked with entries of the OD. The process variables are also used to realize communication in the isochronous phase of the EPL cycle through the PRes frames. Two main types of process variables are used in the stack, namely input and output process variables. The former handle the processing as well as manipulation of sensor/actuator data (i.e. ATD conversion, encoding/decoding, data display) and are also initiating data transmission. On the other hand the latter are useful for data reception.

Further techniques in the openPOWERLINK stack is the assignment of higher priority to data handling during the EPL cycle than event handling. This ensures its real-time behavior and is accomplished with the use of threads. Furthermore, it supports real-time communication between modules of one or several layers is the presence of *asynchronous callbacks*. Callbacks are functions that are passed as an argument to another function and are commonly using in event-driven programming. In this way a callback is used to subscribe to an event and accordingly invoked when the event happens. As an example, in openPOWERLINK callbacks are used in the EPL Application for data transmission in which case a callback has to be defined for the communication with the user as well as the kernel part. The most commonly used callbacks in openPOWERLINK define communication between the Communication Abstraction Layer (CAL) and Data Link Layer for the kernel part as well as between the EPL Application and the API layer for the user part. The former is used to update the values of the process variables during the EPL cycle and the latter to provide event-based notifications from the user part modules (i.e Event Handler, OD, PDO, SDO) to the EPL Application.

Until now we have presented techniques which are used in openPOWERLINK to successfully address and handle the aforementioned challenges. The described techniques can be used for the development of functional applications in openPOWERLINK, which is sequential and relies on the following steps.

1. The MN should detect and access the connected CNs in the EPL network through an Ident Request
2. The Object Dictionary entries in the CNs are initialized by dedicated SDO frames in the asynchronous phase of the EPL cycle
3. The process variables of the EPL Application layer should be linked with entries of the OD module for each node (MN or CN). Once linked, a modification of a process variable will automatically signal the API layer to update the dedicated entry in the node's OD.
4. Implementation of the callback between the Communication Abstraction Layer (CAL) and the Data Link Layer for the kernel part.
5. Implementation of the callback between the EPL Application and the API layer for the user part.

1.4.2.12 Application Development for a Controlled Node

The CN should be able to reply in certain time frames whenever polled by the MN device. It is configured according to the following steps:

1. Reply to the MN's Ident Request
2. Link the API process variables to the OD

3. Store the API process variables in the OD
4. Implement the callback function called by the event handling module in the kernel part
5. Implement the callback function called by the API layer in the user part

1.4.2.13 EtherCAT

EtherCAT[16] (Ethernet for Control Automation Technology) is developed by Beckhoff Automation Inc. as an Ethernet-based Fieldbus system. The main objective was using Ethernet in automation applications with short response time ($\leq$100 µs) and low jitter for accurate synchronization ($\leq$1 µs)). To sum up, the EtherCAT protocol is an optimized Ethernet protocol devised for short cyclic process data.

1.4.3 Future Directions: Interaction with the IoT Ecosystem

1.4.3.1 Wireless and Cellular Infrastructure

Industrial devices (e.g. PLCs) can communicate with the Cloud to perform data processing and analysis as well as allow remote maintenance and troubleshooting from operation engineers. Data exchange is can be performed by integrating the industrial devices with solutions that belong in three main categories:

- *5G radio access technologies*: This technology provides wide area, broadband access. The 5G technology is currently in the process of conceptual development and standardization by the World Radiocommunication Conference (WRC). The 5G technology is expected to have a specific V2X aspect of the 5G technology in a practical scale after 2020. However, in this document we are leveraging the limited standardization to illustrate conceptually its main scope and architectural view.
- *Pre-5G radio access technologies*: Multiple cellular technologies were identified by the ETSI 3rd Generation Partnership Project (3GPP), *LoRa* Alliance and other organizations, such as Narrowband IoT [45], Long Term Evolution for Machines (LTE-M) [44], LoRa are [37] considered. Even though these technologies are already used in V2P/P2V, the main challenge when adopting them in other V2X communication types are reliability and safety, which are currently not addressed in the scope of Low-Power Wide Area Networks (LPWAN).
- *Non-cellular technologies providing wireless access*: IEEE has defined different standards for wireless communication, such as 802.11ac and 802.11p, however only 802.11p is flexible in terms of throughput and offers higher reliability, even though its maximal throughout is more limited than 802.11ac (from 3 to 27 Mbps

[16]More information available at: https://www.ethercat.org/en/technology.html.

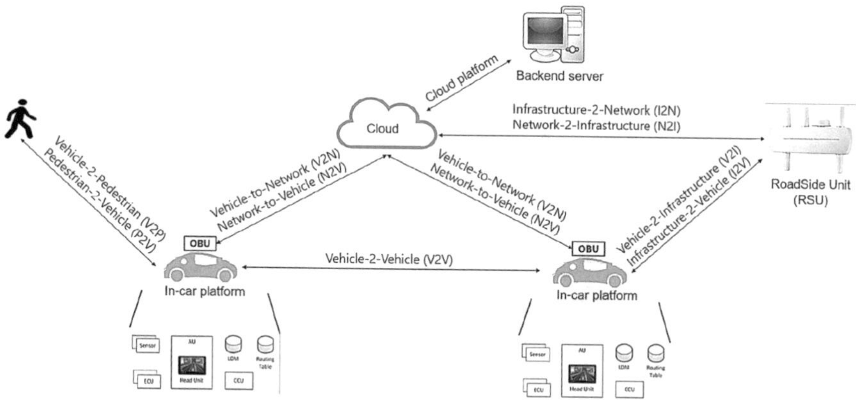

Fig. 1.19 V2X communication types

raw data rate). The reason behind this is that 802.11p was designed particularly for safety-related Vehicular Ad-hoc NETworks (VANET), including the V2V and V2I/I2V concepts. IEEE 802.11p technology is currently fully specified and already deployed in different locations.

The following paragraphs start with a description of the scenarios supported by 802.11p communication and cellular communication. This is followed by a description on both the 802.11p and 5G technologies. In the scope of this section we focus on these two technologies, because, to the best of our knowledge, they are considered as the leading candidates for V2X communication.

1.4.3.2 Connected Cars

Industrial IoT technologies allow vehicles to evolve by the integration of technologies that allow them to take decisions autonomously without any human intervention. This allows vehicles to communicate with many external entities in context that is now referenced as Vehicle to Everything (V2X). V2X enables the following scenarios in vehicles:

- Vehicle-2-Vehicle (V2V)
- Vehicle-2-Infrastructure (V2I)/Infrastructure-2-Vehicle (I2V)
- Vehicle-2-Pedestrian (V2P)/Pedestrian-2-Vehicle (P2V)
- Vehicle-2-Network (V2N)/Network-2-Vehicle (N2V)
- Infrastructure-2-Network (I2N)/Network-2-Infrastructure (N2I)

These types along with their interactions are demonstrated in Fig. 1.19.

The ultimate goal of the integration would be the fleet management by each industrial manufacturer, that would allow remote diagnostics and monitoring for each factory-built vehicle. Specific use-cases for fleet-management are:

- Fault prediction based on analytics
- Remote maintenance
- Energy management.

1.4.3.3 Smart Grid (e.g. DNP3, DLMS/COSEM Protocols)

Industrial IoT technologies bring a revolution also to the traditional power grid. Specifically, the traditional power grid consisted of isolated small systems and was limited to small geographical areas, which led to local power generation, local limited transmission capabilities and local power distribution. Instead by integrating IoT technologies the power grid transitions towards the smart grid, where all the systems are interconnected. To understand more on the interconnection of smart grid systems we illustrate in Fig. 1.20, the four main systems that they are composed of as well as the assets (i.e. devices) that handle data exchange in them [40]. A detailed explanation of the main systems also follows:

- Generation: The utility company has dedicated assets, called generators, to provide substations with enough energy, which they can distribute to the utility substations.
- Transmission: The utility substation of the energy distribution company is responsible for distributing the electricity to entire cities or neighbourhoods. In this system we can also find switching, protection and control equipment, and circuit breakers in order to interrupt any short circuits or overload currents that may occur on the network.
- Distribution: It is handling the dispatch of energy through the Utility Access Points and associated assets, called transformers, responsible for converting high voltage power used in power lines to lower voltage power that can be used safely in residential homes and businesses. This system also includes data concentrators, that are responsible for calculating the difference between the actual energy consumption in each individual house and compares to an estimated consumption that is associated with historical consumption data for each house as well as with the climate conditions in the house area.
- Consumption: This system concerns the usage of electricity in the consumer houses or factory buildings that employ smart appliances and smart meters in order to manage and optimize the energy consumption in the system. Smart meters are an important source of distributed intelligence for the smart grid and hence are often referred to in literature as Advanced Metering Infrastructure (AMI).

If we focus now on the communication protocols that enable the transition to the smart grid we observer three main protocols that are used for data exchange: DLMS/COSEM and DNP3.

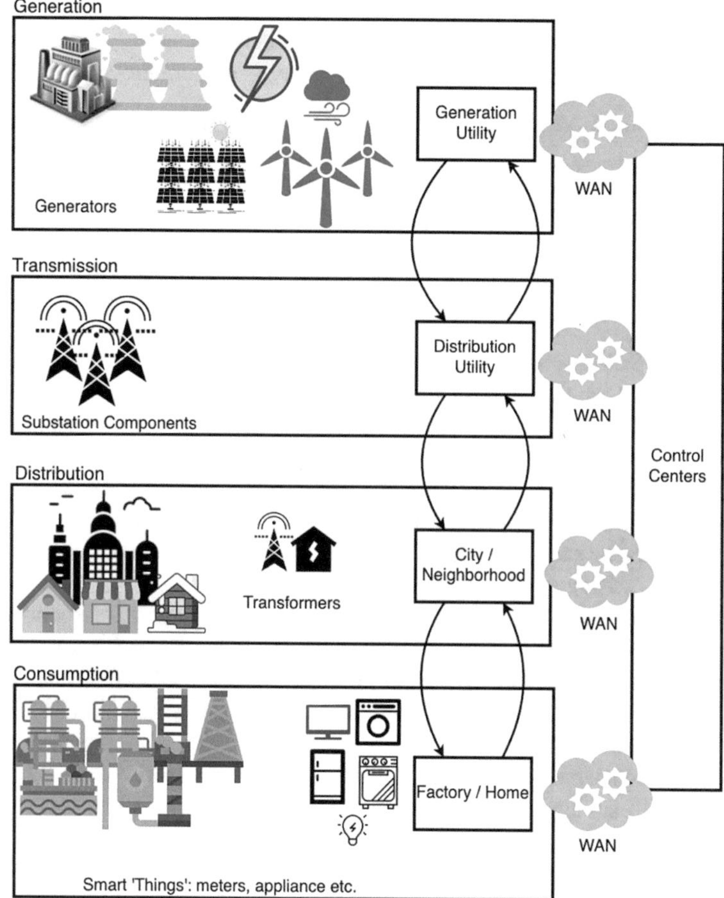

Fig. 1.20 Smart grid architecture based on NIST smart grid conceptual model

DLMS/COSEM

DLMS/COSEM is a widely used communication protocol for metering devices. DLMS stands for "Device Language Message specification" and describes a concept for modeling of communication entities; COSEM stands for "COmpanion Specification for Energy Metering" and sets the rules for data exchange with energy meters. The application fields for DLMS/COSEM are in electricity metering or gas metering within the commercial & industrial environment. Also other applications are feasible, for example heat or water.

DLMS/COSEM represents a client-server architecture. In common operation mode the client is represented by the Head-End System that connects to the metering device, which represents the server. Furthermore, push mechanisms are also available for sending information for the server directly to the client. For example this needed when sending critical alarms from a metering device to the Supervisory Control And Data Acquisition (SCADA) System.

DLMS/COSEM defines the following for smart meter communication:

- An object model, to view the functionality of the meter, as it is seen at its interface(s)
- An identification system for all metering data
- A messaging method to communicate with the model and to turn the data to a series of bytes
- A transporting method to carry the information between the metering equipment and the data collection system

DLMS/COSEM specification covers five ISO/OSI communication layers that are depicted in Fig. 1.21. Specifically, DLMS/COSEM supports both HDLC [19] and IP [20] transport layers. The smart meter software provided supported HDLC, Transmission Control Protocol (TCP), and User Datagram Protocol (UDP) transports.

The P0 optical port (Fig. 1.22) on some smart meters support IEC-62056-21 [21] mode E, or DLMS/COSEM over HDLC. The protocol for requesting the meter to enter mode E is the same as is described in the standard but in practice, timing was an big issue in setting up the physical connection.

Security in DLMS/COSEM is specified through a data protection security layer that provides encryption and authentication mechanisms. Encryption is based on Elliptic curve digital signature (ECDSA) or Elliptic curve Diffie-Hellman (ECDH) key agreement [24]. The authentication mechanism includes three procedures: (1) no security, where no identification takes place, (2) Low Level Security (LLS), where the DLMS server identifies the DLMS client by password and (3) High Level Security (HLS) authentication, where both DLMS server and client and mutually identified by challenge exchange and verification of the challenge result Fig. 1.23 shows the different authentication mechanisms between a DLMS client and a server.

DNP3

Distributed Network Protocol-3 (DNP3) is published by IEEE in 2010 and is a standard based on the IEC 61850 standard, which defines a set of communication protocols used in process automation systems, especially utility distribution, such as electricity and water. The protocol is developed to assess communication in between various types of data acquisition and control equipment, such as SCADA(s), Remote Terminal Units (RTUs), and Intelligent Electronic Devices (IEDs) [14].

1.4.3.4 Building Automation System (e.g. ZigBee, BACnet)

Building Automation (BA) is designed to automate and control buildings‘ air conditioning, elevator, heating, lighting, and other smart devices via Building Automation System (BAS) to improve the comfort and reduce the overall energy

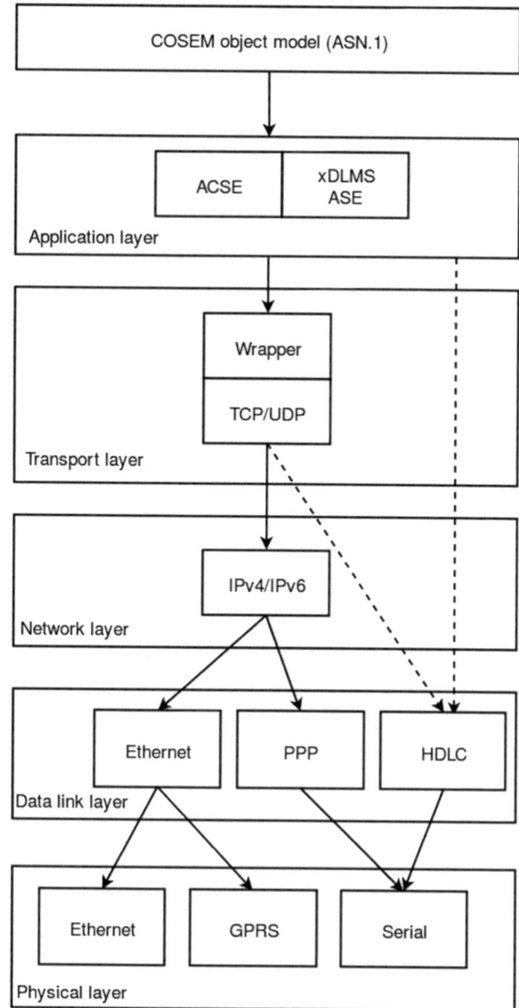

Fig. 1.21 DLMS/COSEM protocol stack

cost. BA is a good example of distributed control systems since there could be multiple gas/smoke detectors, light, security, temperature, water sensor in the building. Fiber, Ethernet or wireless communication is used to provide physical connectivity between the devices. Protocols like BACnet, ZigBee and various other widely used in BA. BACnet (Building Automation and Control Networks) leverages the ASHRAE, ANSI, and ISO 16484-5 standard protocols.

ZigBee is a short-range (10–100 m) low power (low energy consumption) wireless communication protocol based on IEEE 802.15.4 standard.

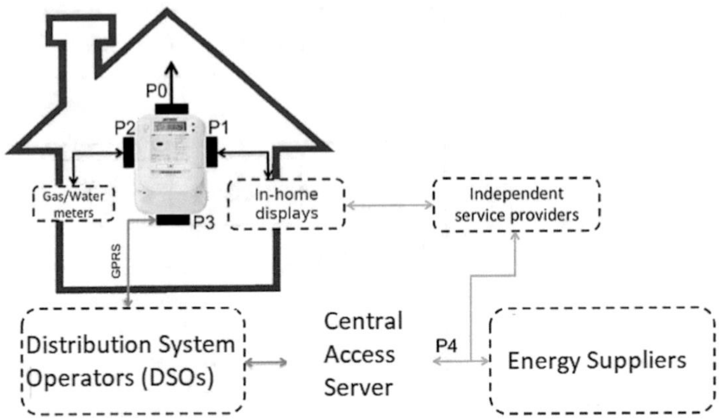

Fig. 1.22 Smart meter ports that generate data

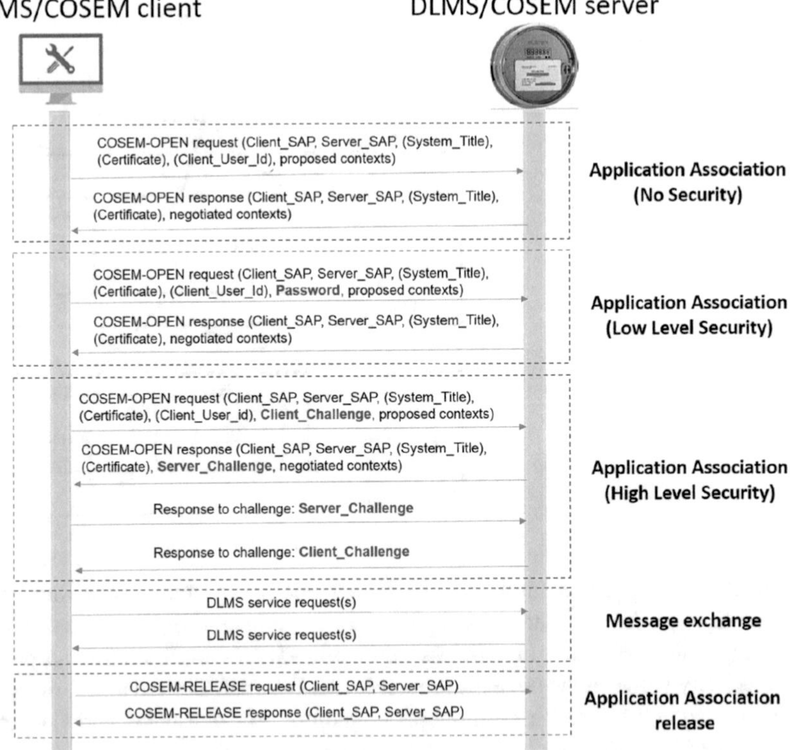

Fig. 1.23 Authentication mechanisms between workstation (DLMS client) and smart meter (DLMS server)

1.5 Conclusions

In particular, following are the main contributions of this chapter:

1. New advancements in the field of industrial networks, especially IIoT are provided.
2. Security related implications and projections are provided.
3. Most of the prominent networking and communication technologies are provided in a comparative way.

Authors of this chapter foresee that future industrial networks, especially their IDSs, will take advantage of data streaming concept in their algorithms, due to several reasons: (1) To provide a rapid response by removing the necessity of the round-trip messages in between the edge of the network and the server, (2) to decrease the amount of traffic created hence causing less bottlenecks in the network, and finally, (3) to reduce the overall response time for the queries by processing the data closer to the source.

As a conclusion, future industrial and IIoT networks will benefit from various emerging advanced technologies such as LoRa, Nb-IoT, 5G, SDN, SDR, Fog/Edge, etc., to be able to implement the needs of the Industry 4.0 such as remote observing, monitoring, and commanding features along with the Digital Twin.

Acknowledgments This research has been partially supported by the Swedish Civil Contingencies Agency (MSB) through the projects RICS, by the EU Horizon 2020 Framework Programme under grant agreement 773717, and by the Swedish Foundation for International Cooperation in Research and Higher Education (STINT) Initiation Grants program under grant agreement IB2019-8185.

References

1. Alley, G.: What Is Data Streaming? https://dzone.com/articles/what-is-data-streaming. Accessed 12 July 2019
2. Axel Pöschmann Lutz Rauchhaupt, T.W.: Integration of CAN-based networks into the PROFInet environment. In: 9th International CAN Conference, Munich, Germany (2003)
3. Aydogan, E., Yilmaz, S., Sen, S., Butun, I., Forsström, S., Gidlund, M.: A central intrusion detection system for rpl-based industrial internet of things. In: 2019 15th IEEE International Workshop on Factory Communication Systems (WFCS), pp. 1–5. IEEE, Piscataway (2019)
4. Baumgartner, J., Schoenegger, S.: Powerlink and real-time linux: a perfect match for highest performance in real applications. In: Twelfth real-time linux workshop, Nairobi, Kenya (2010)
5. Bizanis, N., Kuipers, F.A.: SDN and virtualization solutions for the internet of things: a survey. IEEE Access **4**, 5591–5606 (2016)
6. Butun, I.: Prevention and detection of intrusions in wireless sensor networks. University of South Florida, Ph.D. thesis (2013)
7. Butun, I.: Privacy and trust relations in internet of things from the user point of view. In: 2017 IEEE 7th Annual Computing and Communication Workshop and Conference (CCWC), pp. 1–5. IEEE, Piscataway (2017)

8. Butun, I., Österberg, P.: Detecting intrusions in cyber-physical systems of smart cities: Challenges and directions. In: Secure Cyber-Physical Systems for Smart Cities, pp. 74–102. IGI Global, Hershey (2019)
9. Butun, I., Morgera, S.D., Sankar, R.: A survey of intrusion detection systems in wireless sensor networks. IEEE Commun. Surv. Tutorials **16**(1), 266–282 (2013)
10. Butun, I., Pereira, N., Gidlund, M.: Analysis of lorawan v1. 1 security. In: Proceedings of the 4th ACM MobiHoc Workshop on Experiences with the Design and Implementation of Smart Objects, p. 5. ACM, New York (2018)
11. Butun, I., Österberg, P., Song, H.: Security of the internet of things: vulnerabilities, attacks and countermeasures. IEEE Commun. Surv. Tutorials (2019). https://doi.org/10.1109/COMST.2019.2953364
12. Butun, I., Pereira, N., Gidlund, M.: Security risk analysis of lorawan and future directions. Future Internet **11**(1), 3 (2019)
13. Butun, I., Sari, A., Österberg, P.: Security implications of fog computing on the internet of things. In: 2019 IEEE International Conference on Consumer Electronics (ICCE), pp. 1–6. IEEE, Piscataway (2019)
14. Butun, I., Lekidis, A., dos Santos, D.: Security and privacy in smart grids: challenges, current solutions and future opportunities. In: Proceedings of the 10th International Conference on Information Systems Security and Privacy (ICISSP), pp. 1–6. Springer, New York (2020)
15. CAN in Automation: Application Note 802 (2005)
16. CAN in Automation: Electronic data sheet specification for CANopen, Draft Standard 306 (2005)
17. CAN in Automation: CANopen device description, Draft Standard 311 (2007)
18. CAN in Automation: Application layer and communication profile, Draft Standard 301 (2011)
19. Commission, I.E., et al.: Electricity metering - data exchange for meter reading, tariff and load control. IEC-62056 Part 46 **21**, 31–51 (2002)
20. Commission, I.E., et al.: Electricity metering - data exchange for meter reading, tariff and load control. IEC-62056 Part 47 **21**, 31–51 (2002)
21. Commission, I.E., et al.: Electricity metering data exchange - the dlms/cosem suite-part 5-3: DLMS/COSEM application layer. Tech. rep., IEC 62056-5-3:2016, ed (2016)
22. Decotignie, J.D.: Ethernet-based real-time and industrial communications. Proc. IEEE **93**(6), 1102–1117 (2005)
23. Derhamy, H., Eliasson, J., Delsing, J.: IOT interoperability-on-demand and low latency transparent multiprotocol translator. IEEE Internet Things J. **4**(5), 1754–1763 (2017)
24. DLMS, U.: DLMS/COSEM Architecture and Protocols. Green Book (2019)
25. Eldefrawy, M., Butun, I., Pereira, N., Gidlund, M.: Formal security analysis of lorawan. Comput. Netw. **148**, 328–339 (2019)
26. Farris, I., Taleb, T., Khettab, Y., Song, J.: A survey on emerging sdn and nfv security mechanisms for iot systems. IEEE Commun. Surv. Tutorials **21**(1), 812–837 (2018)
27. Farwell, J., Rohozinski, R.: Stuxnet and the future of cyber war. Survival **53**, 23–40 (2011). https://doi.org/10.1080/00396338.2011.555586
28. Ferrari, P., Sisinni, E., Brandão, D., Rocha, M.: Evaluation of communication latency in industrial IOT applications. In: 2017 IEEE International Workshop on Measurement and Networking (M&N), pp. 1–6. IEEE, Piscataway (2017)
29. Grzywaczewski, A.: Training ai for self-driving vehicles: the challenge of scale (2019). https://devblogs.nvidia.com/training-self-driving-vehicles-challenge-scale/
30. Halpern, M.: Iran flexes its power by transporting turkey to the stone age. Observer (2015). https://observer.com/2015/04/iran-flexes-its-power-by-transporting-turkey-to-the-stone-ages/
31. Hatzivasilis, G., Askoxylakis, I., Alexandris, G., Anicic, D., Bröring, A., Kulkarni, V., Fysarakis, K., Spanoudakis, G.: The interoperability of things: Interoperable solutions as an enabler for iot and web 3.0. In: 2018 IEEE 23rd International Workshop on Computer Aided Modeling and Design of Communication Links and Networks (CAMAD), pp. 1–7. IEEE, Piscataway (2018)

32. Industrial IoT: Rise of Digital Twin in Manufacturing Sector. https://www.biz4intellia.com/blog/rise-of-digital-twin-in-manufacturing-sector/. Accessed 26 Jan 2020
33. Jeschke, S., Brecher, C., Meisen, T., Özdemir, D., Eschert, T.: Industrial internet of things and cyber manufacturing systems. In: Industrial Internet of Things, pp. 3–19. Springer, Cham (2017)
34. Koc, A.T., Jha, S.C., Gupta, M., Vannithamby, R.: Extended discontinuous reception (drx) cycle length in wireless communication networks. U.S. Patent Application No 14 (2013)
35. Kocakulak, M., Butun, I.: An overview of wireless sensor networks towards internet of things. In: 2017 IEEE 7th Annual Computing and Communication Workshop and Conference (CCWC), pp. 1–6. IEEE, Piscataway (2017)
36. Lekidis, A.: Design flow for the rigorous development of networked embedded systems. Ph.D. thesis (2015)
37. LoRa Alliance: A technical overview of lora and lorawan (2015)
38. Measuring the Information Society Report. https://www.itu.int/en/ITU-D/Statistics/Documents/bigdata/MIS2015-Chapter5.pdf. Accessed 26 Jan 2020
39. Mitola, J.: Software radios: survey, critical evaluation and future directions. IEEE Aerospace Electron. Syst. Mag. **8**(4), 25–36 (1993)
40. Mo, Y., Kim, T.H.J., Brancik, K., Dickinson, D., Lee, H., Perrig, A., Sinopoli, B.: Cyber-physical security of a smart grid infrastructure. Proc. IEEE **100**(1), 195–209 (2011)
41. Najdataei, H.: Parallel data streaming analytics in the context of internet of things. Licentiate Thesis, Chalmers University of Technology (2019)
42. Pfeiffer, O., Ayre, A., Keydel, C.: Embedded Networking with CAN and CANopen. Copperhill Media, Greenfield (2008)
43. Prytz, G.: A performance analysis of ethercat and profinet IRT. In: IEEE International Conference on Emerging Technologies and Factory Automation, 2008. ETFA 2008, pp. 408–415. IEEE, Piscataway (2008)
44. Ratasuk, R., Mangalvedhe, N., Ghosh, A., Vejlgaard, B.: Narrowband LTE-M system for M2M communication. In: Vehicular Technology Conference (VTC Fall). IEEE, Piscataway (2014)
45. Ratasuk, R., Mangalvedhe, N., Zhang, Y., Robert, M., Koskinen, J.P.: Overview of narrowband iot in lte rel-13. In: IEEE Conference on Standards for Communications and Networking (CSCN) (2016)
46. Sagiroglu, S., Terzi, R., Canbay, Y., Colak, I.: Big data issues in smart grid systems. In: 2016 IEEE International Conference on Renewable Energy Research and Applications (ICRERA), pp. 1007–1012. IEEE, Piscataway (2016)
47. Std., I.E.C.: IEC 61784: Digital data communications for measurement and control - part 1: Industrial communication networks - profiles - part 1: Fieldbus profiles (August 2014)
48. Std., I.E.C.: IEC 61784: Digital data communications for measurement and control - part 2: Additional profiles for ISO/IEC8802-3 based communication networks in real-time applications (July 2014)
49. Steer, M.: Will There Be A Digital Twin For Everything And Everyone? www.digitalistmag.com. Accessed 08 Oct 2018
50. Steiner, W.: Ttethernet specification. TTTech Computertechnik AG (Nov 2008)
51. Tao, F., Cheng, J., Qi, Q., Zhang, M., Zhang, H., Sui, F.: Digital twin-driven product design, manufacturing and service with big data. Int. J. Adv. Manuf. Technol. **94**(9–12), 3563–3576 (2018)
52. Tello-Oquendo, L., Akyildiz, I.F., Lin, S.C., Pla, V.: Sdn-based architecture for providing reliable internet of things connectivity in 5g systems. In: 2018 17th Annual Mediterranean Ad Hoc Networking Workshop (Med-Hoc-Net), pp. 1–8. IEEE, Piscataway (2018)
53. Zeltwanger, H.: Gateway profiles connecting CANopen and Ethernet. In: 10th International CAN Conference, Rome, Italy (2005)
54. Zhang, Y., Huang, T., Bompard, E.F.: Big data analytics in smart grids: a review. Energy Inform. **1**(1), 8 (2018)

Chapter 2
Wireless Communication for the Industrial IoT

Hasan Basri Celebi, Antonios Pitarokoilis, and Mikael Skoglund

2.1 Introduction

Transmission of information from a source to the receiver over a distance without using cables or any other transmission wires is called wireless communication. Wireless communication is a general concept that covers all methods of linking and communicating between two or more devices through a wireless channel.

Today, a tremendous number of communication systems are wireless and this number is growing every day because of the advantages and business benefits that wireless communication systems can offer [22]. Wireless technologies offer cost effectiveness, mobility, flexibility, connectivity, and speed [49]. Since information is transmitted over the air, wireless connectivity is counted as a cost-effective way to provide communication coverage to wide areas and therefore wireless networks are more affordable to install and maintain. Transmission over the air also provides the freedom of mobility which increase the flexibility and connectivity of the communication system. With the latest improvements on the wireless technologies, nowadays, they offer data transmission speeds comparable to their wired competitors.

However, wireless communication also has its own disadvantages. For instance, if not secured properly, the wireless medium is open to security threats [3]. Wireless communication spectrum is a limited resource and therefore has to be carefully distributed over various technologies, applications, and companies [65]. On the other hand, the wireless communication channel is a particularly hostile communication medium [49]. A signal is reflected, scattered and dispersed by various objects on its way from the source to an intended destination, which can

H. B. Celebi (✉) · A. Pitarokoilis · M. Skoglund
Electrical Engineering and Computer Science Department, KTH Royal Institute of Technology, Stockholm, Sweden
e-mail: hbcelebi@kth.se; apit@kth.se; skoglund@kth.se

I. Butun (ed.), *Industrial IoT*, https://doi.org/10.1007/978-3-030-42500-5_2

lead to severe attenuation of the received energy and substantial distortion of its spectral properties. Further, since the environment can be rapidly changing, the received signal is subject to considerable fading of its power due to destructive interference arising from multipath [23] propagation or due to objects blocking the signal propagation. Ensuring levels of communication reliability, rate of information transmission and latency that are comparable to more controlled communication mediums, such as wired communication, is essential for the successful realization of the services that IoT promises.

In this chapter, wireless communication for the industrial IoT (IIoT) applications is discussed. First, state-of-the-art wireless standards and technologies are presented in terms of suitability for IIoT applications. Second, challenges for mission-critical IIoT applications are studied. Therefore, initially, theoretical limits on the capacity of wireless channels are discussed, and then we focus on ultra-reliable low-latency communication (URLLC), which is one of the most significant technologies that is crucial for mission-critical applications. Thus, low-latency communication with complexity constraint receivers is studied, where interesting trade-offs are revealed for complexity constrained IIoT nodes when a constraint on transmission and decoding latency is applied. Understanding these trade-offs is crucial in order to design a reliable communication system for mission-critical IIoT applications.

2.2 Wireless Communication for IIoT Applications

Wireless connectivity is an indispensable component of IoT infrastructure. IoT must enable devices with vastly different characteristics and requirements to communicate seamlessly, over heterogeneous networks, without human supervision. The need for autonomous operation of IoT devices implies that they should be capable of functioning without being attached to any fixed wired or optical fiber infrastructure.

Thanks to the enhancements in wireless communication technologies and proliferation of connected devices such as mobile phones, we are living the digital world. But still the IoT and IIoT are intended to broaden further the Internet into the physical world by enabling seamless communication environment between physical objects in our daily lives and in industrial applications, respectively. In general, IoT is an umbrella term to cover the operation of low-power and low-complexity communication devices to be applied in various fields such as wearable sensors, smart-home appliances, smart-transportation, smart-grid and smart industrial applications [37].

Each field has different application areas which have their own characteristics. Therefore, for each field, latency, data rate, and reliability requirements may change, and yet currently there is no all-in-one wireless communication

(continued)

standard that appropriately fits as a global solution [58]. For instance, battery-powered sensor node IIoT devices are expected to be functional and reliable for a number of years, therefore energy efficiency is crucial, whereas critical IIoT devices, such as real-time emergency monitoring and safety traffic management, require low-latency and ultra-reliability features in order to guarantee the end-user quality of service since an error in such applications may result in severe consequences. Therefore, providing the required seamless communication between "things", is a challenging endeavor that calls for the judicious incorporation of the existing standards as well as the invention of new approaches that are tailored to the needs of each application [2].

Latency and reliability requirements of some important IIoT applications are listed below and in Table 2.1, where the traffic types are sorted in least critical to most critical order.

- **Monitoring traffic**: The purpose of this type of traffic is to gather information from systems such as reading non-critical data logging from IIoT sensors in control or automation. Therefore, data transmission can span from minutes to hours and since packet loss is common, low reliability is acceptable.
- **Non-critical alerting**: In general, non-critical alerting traffics consist of regular feedback data from industrial processes, such as non-critical temperature readings.
- **Critical alerting**: Critical alerting traffics require periodic controls from the sensors. Medium latency and reliability values are acceptable.
- **Remote control traffic**: Remote control for autonomous vehicles is important in future industrial applications. Latency and reliability values up to 50 ms and $1 - 10^{-3}$ are required, respectively [53].
- **Critical control traffic**: Critical control traffic is required to ensure a smooth flow of industrial processes such as robotic controls. Low latency values up to 1 ms with codeword error rates (CER) not more than 10^{-8} are required [63].
- **Emergency traffic**: Safety in industrial environments is always very important. Leakage of poisonous gases or emission of radiation need to be alerted immediately. Even a small delay on packet transmission may cause severe results.

Table 2.1 Overview of typical requirements of some IIoT applications [2, 50]

Criticality	Category	Application	Latency	Reliability
Least	Monitoring	Static feedback	Minutes	Low
	Non-critical alerting	Temperature measurement	Minutes	Low
↓	Critical alerting	Periodic control checks	Seconds	Medium
	Remote control traffic	Unmanned vehicle control	Miliseconds	High
	Critical control traffic	Robotic motion control	Miliseconds	Ultra-high
Most	Emergency traffic	Safety traffic	Sub-miliseconds	Ultra-high

Therefore, the communication system that carries emergency traffic needs to support sub-milisecond latency with ultra-high reliability.

As listed above, requirements change drastically from one industrial application to another. With an increment in the level of criticality, lower latency values with higher reliabilities are required. Therefore, in order to satisfy stringent requests of these applications, variety of different wireless resources need to be investigated.

2.2.1 State-of-the-Art Wireless Standards for IIoT Applications

Wireless communication gained extensive popularity in the last decades. Nowadays, wireless communication technologies are accepted to be the primary solution for many industrial applications. This is achieved after many significant improvements in both the physical and MAC layers. Today, there are several standards which are suitable for IIoT deployments of wireless communication. Significant ones, which are determined by IEEE, are discussed in this section.

- **IEEE 802.11**: IEEE 802.11 is a set of standards for wireless local area networks [28]. Although there are several different branches and versions of IEEE 802.11, the most common ones usually designate the protocols of what most people know as Wi-Fi. Specifically, this technology is originated from the need of wireless connection that can provide robust and high data rate point-to-point or point-to-multipoint communication for urban areas. However, recently, new developments in this standard lead to new versions that support industrial applications [1].
  - **IEEE 802.11af**: This standard, referred as White-Fi, specializes in creating a suitable wireless protocol for industrial applications by using the recently freed frequencies of broadcast television coverage [33]. These sub-1 GHz frequency bands lead to higher coverage and less power consumption which is crucial for low-power IIoT devices.
  - **IEEE 802.11ah**: Similar to IEEE 802.11af, the main goal of the IEEE 802.11ah standard, also known as HaLow, is low-power consumption for the IIoT nodes. This standard uses industrial-scientific-medical (ISM) frequency bands below 1 GHz and can support a massive number of possible nodes [16]. Short data packets with shortened packet loads are transmitted in order to minimize the transmit time and power usage and maximize battery life.
- **IEEE 802.15**: Wireless technologies that fall in the domain of wireless personal area network are standardized in IEEE working group 802.15 [57]. In general, low-power and low data rate industrial applications suits better in this standard. There are ten major working groups under 802.15. Here, we will only focus on the ones which have a better support on industrial applications.

- **IEEE 802.15.1**: This standard covers wireless connectivity with portable and mobile devices within personal operating space. Bluetooth is one of the leading technologies that is based on IEEE 802.15.1.
- **IEEE 802.15.4**: Low data rate, low-power, and low-complexity wireless personal area applications are standardized with this standard [50]. There are several applications run over this standard, such as ZigBee, WirelessHART, and 6LoWPAN.
- **IEEE 802.15.4e**: This standardization working group intends to improve and add functionalities to IEEE 802.15.4 to make it more suitable for industrial needs [12]. IEEE 802.15.4e mainly targets the mission-critical industrial applications which require low-latency and high-reliability constraints.

Note that there are other standards apart from the ones listed above, such as ultra-wideband standardizations. Here, we only listed some of the most common ones referred as a possible solution for IIoT applications. Next, various technologies, that are based on these standardization activities, are presented.

2.2.2 *Potential Wireless Solutions for IIoT*

Although there are several different standardizations, most of the industrial applications nowadays are based on IEEE 802.15.4 and IEEE 802.15.4e. Some significant ones, for which the release dates are shown in Fig. 2.1, are listed below.

- **Wi-Fi**: Wi-Fi technology is based on IEEE 802.11.x standards. Up to now, there are 6 main Wi-Fi generations, e.g. (*i*) 1st generation: 802.11b, (*i*) 2nd generation: 802.11a, (*i*) 3rd generation: 802.11g, (*i*) 4th generation: 802.11n, (*i*) 5th generation: 802.11ac, (*i*) 6th generation: 802.11ax [27]. Every new generation has high interoperability with the previous generations. In general, Wi-Fi uses mid-range spread spectrum with orthogonal frequency division multiplexing technology and offers high data-rate connection with relatively higher power consumption and higher cost on transceiver modules, compared to IEEE 802.15 based solutions. Of course, more power consumption degrades the life-time of battery-powered transmitters which is not desirable for most of the industrial applications. In order to deal with these problems, new IEEE solutions such as White-Fi and Hallow are developed [1].

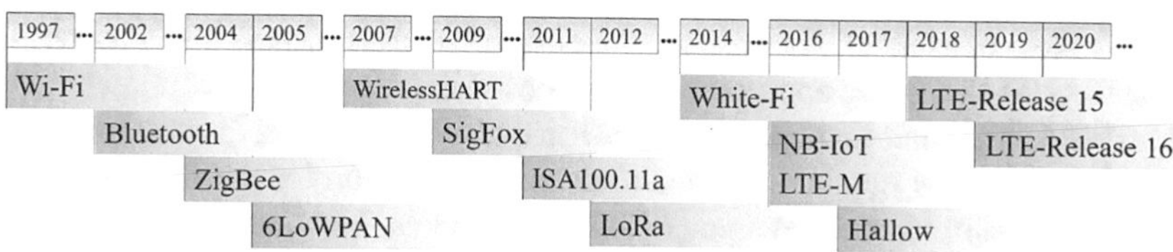

Fig. 2.1 The release timeline of some of the industrial wireless communication technologies

- **Bluetooth**: The main goal of this IEEE 802.15.1 based technology is to provide low-cost and energy-efficient transceiver modules to the market. Today, millions of Bluetooth devices are being used and it is expected that more than 5 billion Bluetooth devices will be sold by 2022 [48]. Bluetooth uses universal short-range radio link with frequency hopping spread spectrum technology which makes it highly immune to interference and increases the security. However, up to now, the main focus of Bluetooth technology was human-machine interfaces but not industrial applications. Recently, numerous number of researchers are proposing improvements and new features on the current Bluetooth protocol in order to make it compliant to the stringent requirements in industrial applications [34, 45].
- **ZigBee**: ZigBee is based on the IEEE 802.15.4 standard and proposed by ZigBee Alliance, which is a non-profit open-source community group. Today, with more than 70 million ZigBee devices operating worldwide, it is one of the most widely used wireless communication solutions. ZigBee mainly focuses on low-power, low-cost, and low data-rate transmission scenarios which suit in most of the industrial applications [5]. However, ZigBee uses carrier sense multiple access schemes which is not suitable for mission-critical industrial applications since it does not offer a consistent solution for high reliability and low end-to-end latency constraints [50].
- **WirelessHART**: This protocol is based on the IEEE 802.15.4 standard and is the wireless version of the wired Highway Addressable Remote Transducer (HART) protocol. WirelessHART was originally developed by HART Communication Foundations in 2007. Today there are more than 40 million industrial devices worldwide that are using WirelessHART, specifically for industrial measurements and control applications. WirelessHART offers high reliability, up to 10^{-5} error rate, and high security by implementing time synched time division multiple accessing technology [24]. However, no stringent constraint on end-to-end latency is introduced. On the other hand, since WirelessHART does not support the Internet protocol (IP), a significant interoperability problem with the next-generation communication systems for IIoT applications may arise.
- **ISA100.11a**: Similar to WirelessHART, ISA100.11a is also based on the IEEE 802.15.4 standard and uses time synched time division multiple accessing in the physical layer to increase reliability and security [46]. ISA100.11a is designated to provide reliable and secure transmission for non-critical monitoring, alerting, open and closed-loop control applications with end-to-end latency values in the order of 100 ms [52]. WirelessHART and ISA100.11a are very suitable for several industrial applications which does not require ultra low-latency. However, interoperability issues with the current IP protocols stands as a serious problem that need to be tackled. Nevertheless, to ensure seamless connection between IP and ISA100.11a, novel approaches are proposed in [52].
- **6LoWPAN**: Similar to the technologies listed above, 6LoWPAN is also based on the IEEE 802.15.4 standard. The main advantage of this protocol is that it enables the use of current and next-generation IP protocols, simultaneously [26]. By communicating over IP networks, 6LoWPAN allows the communication system to directly connect to the other networks simply by connecting to an IP router,

which significantly decreases the implementation cost and makes it prominent for various IIoT applications. 6LoWPAN is capable of meeting specific latency requirements as low as few miliseconds. However, this can only be supported in unreliable mode [42].

- **LoRa**: Based on the media access protocol LoRaWAN, LoRa is a physical layer protocol for low-power wide area networks. LoRa is a convenient and promising technology for low-cost, low-power and long-range wireless networks. Similar to Bluetooth, LoRa also uses frequency hopping spread spectrum technology. On the other hand, LoRa uses unlicensed sub-1GHz bands to enable better coverage in the order of several kms in urban environments with data-rates up to 50 kbps [64].
- **SigFox**: Similar to LoRa, SigFox is also a low-power wide area network solution for low-power, low-cost and long-range requirements. It is based on a patented ultra-narrowband technology and uses sub-1 GHz bands. SiGFox enables the connection of massive number of devices with ranges up to 20 km in rural areas. A fixed packet size, e.g. 12 bytes, is transmitted successively three times in order to increase reliability. However, end-to-end latency values on the order of seconds are allowed [64]. SigFox and LoRa are suitable for most of the monitoring applications, such as water-meter reading, but not for mission-critical applications due to the high end-to-end latency values.

Table 2.2 summarizes and gives some numerical information about the existing wireless IIoT technologies. In addition to the specified standards and technologies, the 3rd Generation Partnership Project (3GPP) has started to work on the standardization of the next generation of mobile wireless communications, namely 5G, and listed the requirements of several emerging applications.

Table 2.2 Summary of the current wireless IIoT technologies [2, 50]

Technology	Frequency band	Max. data rate	Coverage	Latency	Standard
Bluetooth	2.4 GHz	2 Mbps	50 m	200–500 ms	IEEE 802.15.1
White-Fi	Sub-1 GHz	24 Mbps	3 km	–	IEEE 802.11.af
Halow	Sub-1 GHz	100 Mbps	1 km	–	IEEE 802.11.ah
Wi-Fi	2.4 and 5 GHz	54 Mbps	100 m	20–250 ms	IEEE 802.11a
ZigBee	Sub-1 and 2.4 GHz	167 kbps	100 m	60–150 ms	IEEE 802.15.4
WirelessHART	Sub-1 and 2.4 GHz	250 kbps	225 m	0.5–10 s	IEEE 802.15.4
ISA100.11a	Sub-1 and 2.4 GHz	250 kbps	225 m	100 ms	IEEE 802.15.4
6LoWPAN	Sub-1 and 2.4 GHz	250 kbps	100 m	2–6 ms	IEEE 802.15.4
LoRa	Sub-1 GHz	50 kbps	15 km	600 ms	LoRaWAN
SigFox	Sub-1 GHz	100 bps	20 km	1–20 s	Patented
LTE-M	Licensed	1 Mbps	50 km	50–100 ms	LTE release 12
NB-IoT	Licensed	50 kbps	40 km	1–10 s	LTE release 13

2.2.3 5G: The Next Generation of Mobile Wireless Communications

Next-generation wireless networks dramatically change the approach to most of the communication services. Up to now, the goal of these services was mainly human-to-human or human-to-machine interaction. However, with the latest trends in industrial evolution, e.g., Industry 4.0, machine-to-machine type of communication is gaining significant importance [13].

Main usage scenarios of the next generation communication technologies, as shown in Fig. 2.2, are listed as: (1) enhanced Mobile Broadband (eMBB), (2) massive Machine-Type Communication (mMTC), and (3) Ultra-Reliable Low-Latency Communication (URLLC) [50].

2.2.3.1 Enhanced Mobile Broadband

This service enables communication with extremely high data rates across a wide coverage area. The goal is to provide better user experience than the current mobile broadband services to enable emerging services such as UltraHD, 4K or 360° live video streaming, virtual reality media and applications, etc.

eMBB is planned to be the first commercialized technology of 5G and will be the extension of the current 4G cellular technologies. In addition to the existing services, eMBB will be capable of delivering higher capacity, e.g. minimum data rate support at 100 Mbps but can be increased up to 10 Gbps, lower latency values

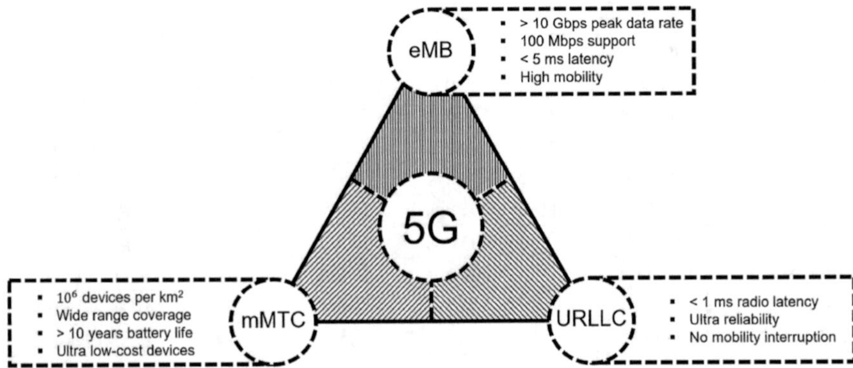

Fig. 2.2 Three main scenarios and their requirements of the next generation 5G wireless communication

around few miliseconds to support virtual reality applications, and high mobility up to 500 km/h in high-speed trains and 1000 km/h in airplanes [40].

2.2.3.2 Massive Machine-Type Communication

The purpose of mMTC is to enable massive number of devices to communicate with the network infrastructure and between each other without any human interaction. mMTC deals with the connectivity of a massive number of devices, in the order of 10^6 devices per km^2 with moderate QoS requirements. This is a challenging goal since, in contrast to classical human-centric communication, e.g., voice traffic, where a constant sampling rate with a constant amount of data is sent, mMTC transmission packets consists of different packet sizes with different QoS requirements and various service features such as (1) low or no mobility, (2) short packet transmission, (3) low power consumption, (4) priority alarm connections [66]. Therefore the following challenges have been identified for mMTC

- Very high connection density: in the order of 10^6 devices per km^2.
- Very low-complexity and low-cost IoT devices and networks.
- Extreme battery life around 10 years.
- QoS requirements with end-to-end latency around 10 s or less depending on the application.

3GPP has been working on machine-type data transmission since 2009 and investigating the requirements. So far, there are two main candidates standardized by 3GPP, listed as (1) Long Term Evolution machine-type communication (LTE-M) and (2) narrow-band IoT (NB-IoT). LTE-M and NB-IoT are specialized to support low power wide area networks with massive number of nodes. Numerical details regarding these two technologies are given in Table 2.2.

2.2.3.3 Ultra-Reliable Low-Latency Communication

URLLC is needed mostly for mission-critical applications and it provides communication support with stringent constraints on reliability and end-to-end latency. In comparison to the aforementioned new services, URLLC is the one which requires the most significant evolution on the current data transmission services since the current technologies are specialized for human centric communication in which end-to-end latency and reliability values are far from the desired targets of URLLC.

2.2.4 Achieving URLLC in IIoT Applications

Industrial applications which require URLLC were first identified by ETSI in [18] in 2011. Based on this technical report, the International Telecommunication Union

(ITU) Radio Communication Sector published a development report in 2015 [29] which lists the requirements and constraints of machine-type communications and URLLC. With the recent major releases, Release-15 and Release-16, the 3GPP Radio Access Network working group aims at listing the official requirements for URLLC and IIoT applications. URLLC is aiming to provide small payloads, around 32–256 bits, with end-to-end radio latency up to 1–10 ms and error probability of less than 10^{-9} [35].

URLLC is crucial for enabling several mission-critical services, such as remote surgery, augmented reality, vehicle automation, industrial robotics, factory automation, smart-grid. Use case requirements of URLLC for some industrial applications are listed in Table 2.3. As Table 2.3 is compared with Table 2.2, it is seen that for most of the industrial URLLC applications there is still no state-of-the-art solution that is eligible to satisfy the requirements. Therefore, research is still ongoing on URLLC [7, 44, 60, and references therein].

For a typical communication application, the end-to-end latency of a packet transmission can be split into several components such as:

- d_t: transmission delay of a single codeword with n symbols.
- d_q: queueing delay occurring in the buffer of access points.
- d_n: networking delay while transmitting the packet throughout different networks.
- d_p: processing delay due to the heavy computation of encoding, decoding, channel estimation, time and frequency synchronization.
- d_l: propagation delay which is caused by the distance between two transceivers.

Suppose that maximum end-to-end delay constraint d_m is introduced to the system. Then, it must hold that

$$d_t + d_q + d_n + d_p + d_l \leq d_m. \tag{2.1}$$

Table 2.3 Overview of typical use case requirements of some URLLC applications [60]

Use case	Latency	Reliability	Data size	Range
Industrial automation	0.25–10 ms	10^{-9}	10–300 bytes	50–100 m
Process automation	50–100 ms	10^{-4}	40–100 bytes	100–500 m
Self driving car	1 ms	10^{-2}	144 bytes	400 m
Smart grid	3–20 ms	10^{-5}	80–1000 bytes	10–1000 m
Tactile internet	1 ms	10^{-7}	250 bytes	10 km

In general, for most of the industrial applications, where the transceivers are not far from each other, d_l is negligible. It is shown in [32] that d_q may have a significant impact on the total latency. Therefore, novel scheduling algorithms are proposed in [4, 15, 32]. Theoretical analyses on delays that are occurred due to queuing are deeply investigated in [54]. Networking delays are analyzed in [14, 43] and references therein and application of software-defined networks are discussed.

Here, the focus will be on the latency components d_t and d_p, which appear to be the most relevant ones for latency constrained communication systems with computationally constrained receivers. Although the chapter mainly investigates channel coding schemes, studies that focus on channel estimation [9, 30], synchronization [31, 36], and resource scheduling schemes [11] for complexity constrained IoT transceivers can also be found in the literature.

2.3 Fundamental Limits on Information Transmission

So far, several state-of-the-art standards and technologies, which are based on excessive research on the limits of the wireless communication systems, have been identified and investigated. We devote the rest of the chapter on challenges of enabling URLLC for IIoT systems with complexity constrained receivers such as mission-critical IIoT receivers. However, in order to accurately model the limits of a communication system with computational complexity constraints, theories need to be revisited. Therefore, we start by introducing the basic theoretical concepts on the asymptotic limits of ultimate transmission rate of a wireless communication system. Here, the limits are derived by assuming that the blocklength of a codeword is infinite, i.e. $n \to \infty$. However, this assumption does not hold for URLLC since stringent constraints on total latency yields transmission of codewords with short blocklengths. As the next step, non-asymptotic limits on transmission rate, where n is finite, are presented. These analyses help us to understand and determine the asymptotic and non-asymptotic information theoretic limits of a communication system.

No constraint on computational complexity of a receiver is taken into account while deriving the theoretic bounds. Therefore, we investigate these limits in case of having constraints on computational complexity. Finally, results are presented and compared with the asymptotic and non-asymptotic information theoretic limits and some interesting trade-offs that appear when enabling URLLC for receivers with computational complexity constraints are shown.

2.3.1 *Communication Channel*

The goal of a communication system is to transfer an information message from the source of the information to a desired destination at a remote location. The source and the destination are connected via a communication channel which can be a wire, an optical fiber, the air, which is the main focus in wireless communication, or some other means that can assist the propagation of the information message. At the source, a device called *transmitter* processes the information message so that it is suitable to be sent through the communication channel. At the destination, a device called *receiver* measures the signal that arrives at its input and provides at its output an estimate of the message that was sent by the transmitter. Since the communication channel is random in nature, the transmitted signal is distorted by a series of random phenomena that are collectively termed as *noise*. Noise can cause the receiver to output a message estimate that is different from the actually transmitted message. This is called an error event and it is the task of the designer of the communication system to select the appropriate transmit and receive processing so that the probability of an error event is kept as low as possible.

It is customary in the study of communication systems to assume that the transmitter selects the message to be transmitted from a, possibly very long, list of messages. Suppose that the message with index j is to be transmitted and that it belongs to a discrete finite set $\mathcal{M} = \{1, 2, \cdots, M\}$. This message can be represented by $k = \log_2 M$ binary digits as

$$\boldsymbol{u}^j = \{u_1, u_2, \cdots, u_k\}, \tag{2.2}$$

where $u_i \in \{0, 1\}$. It is the task of the transmitter to transform the message sequence $\boldsymbol{u}^j$ to some other sequence $\boldsymbol{x}^j = \{x_1, x_2, \cdots, x_n\}$ that is suitable for transmission through the channel.[1] This mapping is called encoding and the component that performs the encoding is the encoder. Since the encoder maps k information bits to a sequence of n channel symbols, the rate of the encoding is given by

$$r = \frac{k}{n} = \frac{1}{n} \log_2 M \tag{2.3}$$

bits per channel input symbol, which is more commonly termed as bits per channel use [bpcu].

Due to its random nature, the channel is often modeled as a conditional distribution $p(\boldsymbol{y}|\boldsymbol{x})$, of the output sequence $\boldsymbol{y}$ conditioned on the input sequence $\boldsymbol{x}$. Each input symbol, x_i, of $\boldsymbol{x}$ belongs to the input alphabet, $\mathcal{X}$, and each output symbol, y_i, of $\boldsymbol{y}$ belongs to the output alphabet, $\mathcal{Y}$. At the receiver, the observation sequence $\boldsymbol{y} = \{y_1, y_2, \cdots, y_n\}$ is mapped back to an index $\hat{j} \in \mathcal{M}$, or equivalently to its k-bit sequence representation, by the decoder. An error occurs when the estimated information sequence index $\hat{j}$ is different from the index j of the information sequence that was actually transmitted. The probability of this error event is denoted by

$$\mathbb{P}(j \neq \hat{j}). \tag{2.4}$$

[1]The symbols, x_i, that constitute the sequence $\boldsymbol{x}^j$, which is suitable for transmission through a given channel, are said to belong to the input alphabet of the channel.

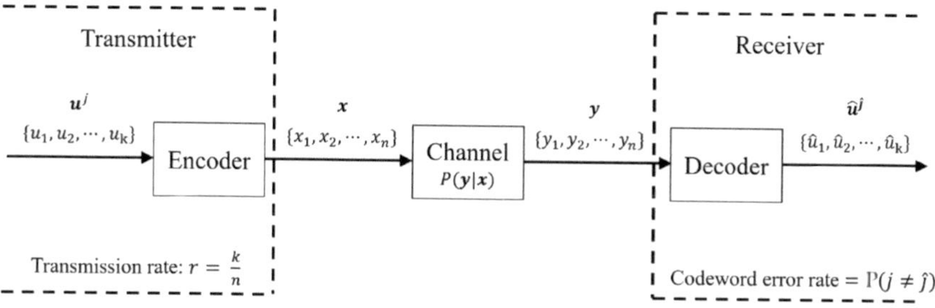

Fig. 2.3 A general block diagram of data transmission

The mapping of the encoder and the demapping of the decoder must be designed so that the probability of an error event is minimized, subject possibly to some additional physical or practical constraints. The basic components in a communication system, described above, are given by the block diagram in Fig. 2.3.

It is clear that the smaller the rate of encoding, r, is the smaller the probability of the error event. When r is small the size of the set of indices $M = \lfloor 2^{nr} \rfloor$ is also small compared to the size of the set of all the possible input sequences of length n generated from the input alphabet $\mathcal{X}$. Consequently, the few message indices can be mapped to input sequences $\boldsymbol{x}$, which, when sent through the channel, generate with high probability output sequences $\boldsymbol{y}$, that are sufficiently different and, hence, unlikely to be confused with each other by the decoder. One would assume that in order to achieve communication with ever decreasing probability of error, the rate r should also decrease. However, in 1948, Shannon in his ground-breaking paper "A Mathematical Theory of Communication" [55], which initiated the field of Information Theory, showed that this is not correct. In fact, we know that for certain channels there exist enconding algorithms with a strictly positive encoding rate, $C > 0$, and corresponding decoding algorithms, such that the probability of error becomes arbitrarily small as the length of the encoding n grows arbitrarily large. This rate is called the channel capacity.

2.3.2 *Channel Capacity*

For the simplicity of the exposition, we limit ourselves to memoryless channels where the input to the channel, x_i, is selected from an arbitrary input space $\mathcal{X}$ and the output, y_i belongs to an arbitrary space $\mathcal{Y}$. We assume that the input sequence $\boldsymbol{x} = \{x_1, x_2, \cdots, x_n\}$ and the output sequence $\boldsymbol{y} = \{y_1, y_2, \cdots, y_n\}$ are related by a conditional probability density function, denoted as

$$p_{\boldsymbol{y}|\boldsymbol{x}}(\boldsymbol{y}|\boldsymbol{x}) = \prod_{i=1}^{n} p_{y|x}(y_i|x_i). \tag{2.5}$$

The factorization of the density function is due to the assumption that the channel is memoryless. No feedback link is assumed to exist between the receiver and the transmitter.

Let $p_y(y)$ be the output density induced by an input density $p_x(x)$. We can define the mutual information density, $\imath(x; y)$, as

$$\imath(x; y) = \log \frac{p_{y|x}(y|x)}{p_y(y)}. \tag{2.6}$$

The ensemble average of the mutual information density, $\imath(x; y)$, with respect to the joint density $p_{y,x}(y, x)$ is the mutual information, $I(x; y)$, given by

$$I(x; y) = \mathbb{E}\left[\imath(x; y)\right]. \tag{2.7}$$

Therefore, $I(x; y)$ depends on the selected input distribution $p_x(x)$. In this case, the channel capacity is given by

$$C = \max_{p_x(x)} I(x; y) \tag{2.8}$$

where the maximum is taken over all allowable input distributions $p_x(x)$.

A memoryless channel with special interest to the following discussion is the binary-input additive white Gaussian noise (Bi-AWGN) channel,

$$y_i = x_i + \omega_i, \quad i = 1, \cdots, n, \tag{2.9}$$

where ω_i is the zero mean additive white Gaussian noise (AWGN) with σ^2 variance: $\omega_i \sim \mathcal{N}(0, \sigma^2)$. For a Bi-AWGN channel, when an input bit 0 is to be transmitted, the symbol $-\sqrt{\rho}$ is sent and when an input bit 1 is to be transmitted, the symbol $\sqrt{\rho}$ is sent. Thus, the input space of the channel is $x_i \in \{-\sqrt{\rho}, \sqrt{\rho}\}$, the output space is $y_i \in \mathbb{R}$ and the conditional probability density function is given by

$$p_{y|x}(y_i|x_i) = \frac{1}{\sqrt{2\pi\sigma^2}} \exp\left(-\frac{\left(y_i + (-1)^{x_i}\sqrt{\rho}\right)^2}{2\sigma^2}\right). \tag{2.10}$$

An illustration of $p_{y|x}(y_i|x_i)$ is depicted in Fig. 2.4.

Without loss of generality, we assume $\sigma^2 = 1$, thereby the signal-to-noise ratio (SNR) of the channel is ρ. Since $p_{y|x}(y_i|x_i)$ depends on ρ, we denote the capacity of the channel as $C(\rho)$ to imply this relation.

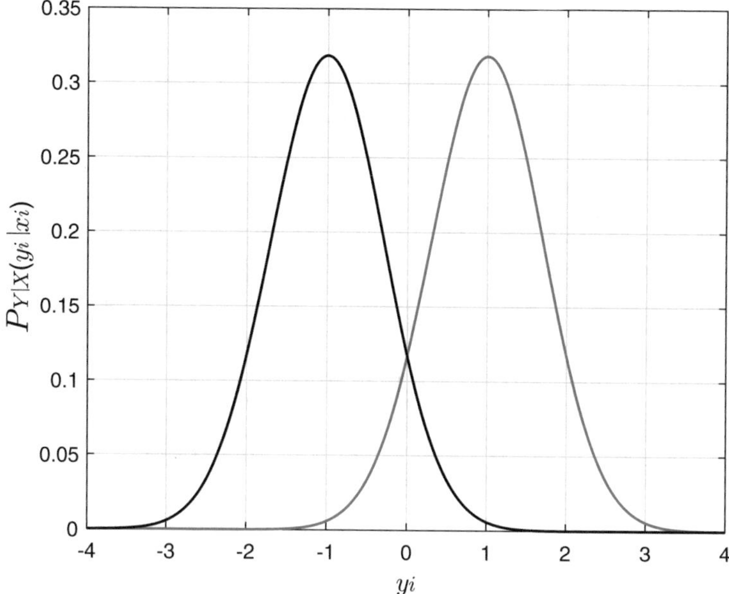

Fig. 2.4 Probability distribution function for y_i where $\rho = 1$, $x_i \in \{-1, +1\}$ and $\sigma^2 = 1$

2.3.3 Finite Blocklength Bounds

It is common in many IIoT communication scenarios to experience sporadic flows organized in packets of few information bits. The channel capacity introduced earlier and other information-theoretic bounds are accurate performance metrics for asymptotically large values of the length, n, of the transmitted sequence $\boldsymbol{x}$. Hence, they often do not represent accurately the performance of communication systems when n is not sufficiently large, such as communication within the IoT setup. Recently, the research area associated with providing information-theoretic metrics that capture more accurately the interplay between the length of transmission blocks, the encoding rate and the probability of error at the receiver, often termed as finite blocklength information theory, has experience a renewed interest within the information and communication theory community [47].

For various channels it has been shown that an approximation of the maximum achievable information rate for messages selected from a codebook of sequences, $\boldsymbol{x}$, with length n channel input symbols and a error probability of ϵ is as follows

$$\mathcal{R}(n, \rho, \epsilon) = C(\rho) - \sqrt{\frac{\mathcal{V}(\rho)}{n}} Q^{-1}(\epsilon) + O\left(\frac{\log n}{n}\right), \tag{2.11}$$

where $C(\rho)$ is the capacity of the channel, $\mathcal{V}(\rho)$ denotes the dispersion of the channel, and $Q^{-1}(\epsilon)$ is the inverse of the Gaussian Q–function

$$Q(z) = \int_z^{\infty} \frac{1}{\sqrt{2\pi}} e^{-\frac{t^2}{2}} \mathrm{d}t. \tag{2.12}$$

A comparison between the capacity, $C(\rho)$, in (2.8), which is asymptotic in the blocklength, n, and the normal approximation for the maximum achievable information rate is shown in Fig. 2.5, where the SNR is fixed to $\rho = 3$ dB and $\epsilon = 10^{-3}$. The dotted line denotes $C(\rho)$ and solid line is the maximum achievable

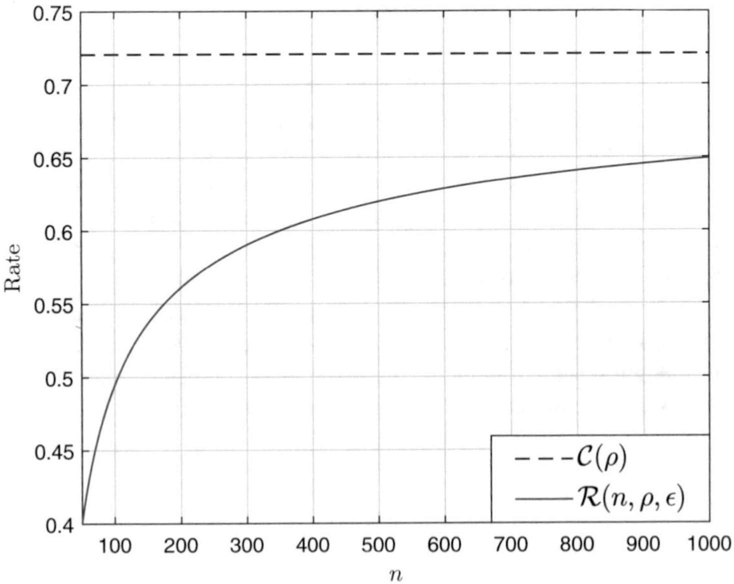

Fig. 2.5 Illustration of the asymptotic capacity (2.8) versus the normal approximation (2.11) for the maximum achievable coding rate for Bi-AWGN channel for different number of channel uses where SNR is $\rho = 3$ dB and $\epsilon = 10^{-3}$

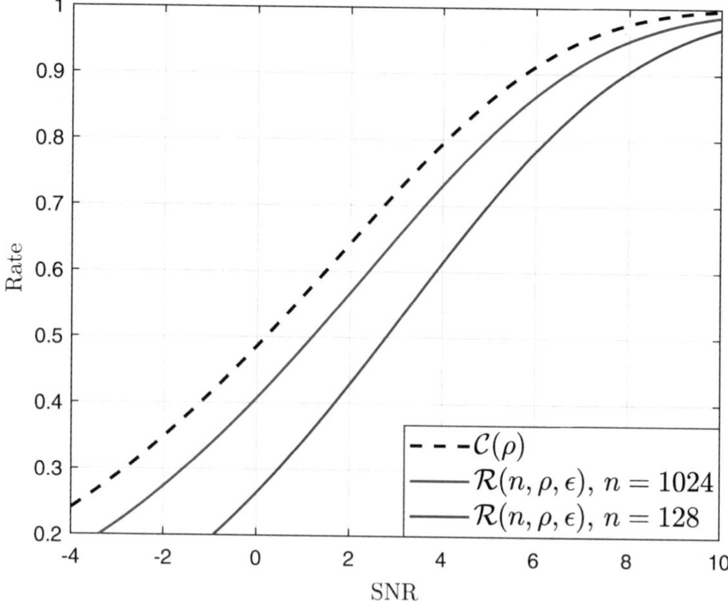

Fig. 2.6 Illustration of the asymptotic capacity (2.8) versus the normal approximation (2.11) for the maximum achievable coding rate for Bi-AWGN channel where $n = \{128, 1024\}$ and $\epsilon = 10^{-3}$

information rate. Since the capacity is an asymptotic in n quantity, it does not depend on n. In contrast, the normal approximation for the maximum achievable information rate increases with n and approaches the capacity as $n \to \infty$.

In Fig. 2.6, the comparison between the capacity and the normal approximation is done with respect to SNR for fixed $n = \{128, 1024\}$ and $\epsilon = 10^{-3}$. Note that, for a fixed SNR, there is a gap between $C(\rho)$ and $\mathcal{R}(n, \rho, \epsilon)$ and this gap increases as n decreases. This gap denotes information rates that achieve reliable transmission when $n \to \infty$ but do not do so when the code blocklength is finite. Therefore, although reliable transmission is possible for any coding rate below $C(\rho)$, there is no coding scheme that can transmit at a rate between $C(\rho)$ and $\mathcal{R}(n, \rho, \epsilon)$ with an error rate less than ϵ for a fixed finite n. Hence, when designing a system with finite-blocklength codes, it is crucial to consider the more elaborate bounds that include the blocklength of the applied code so that a desired reliability can be guaranteed.

2.3.4 *Candidate Channel Coding Schemes for URLLC*

Selection of a channel coding scheme that can achieve the bound is significant in terms of increasing the transmission efficiency of the system. However, there are many coding schemes introduced in the literature. Here, a brief list on some candidate fixed-rate channel codes which are applicable for URLLC is presented.

- **BCH codes with OS Decoder**: The Bose, Chaudhuri, and Hocquenghem (BCH) codes are one of the most powerful error-correcting cyclic codes. BCH codes are independently discovered by Hocquenghem [25] and Bose and Chaudhuri [8] in 1959 and 1960, respectively. BCH codes are flexible codes that can be generated with a wide variety of k and n values. It is shown that there is a binary BCH code with minimum Hamming distance

$$d_{\min} \geq 1 + \frac{2(n-k)}{\log_2(n-1)} \tag{2.13}$$

that can correct up to

$$\frac{n-k}{\log_2(n-1)} \tag{2.14}$$

number of errors, where

$$n = 2^m - 1 \tag{2.15}$$

for some integer m and

$$k \geq n - mt \tag{2.16}$$

where $t < 2^{m-1}$ [51].

Apart from their flexibility on code design and capability of correcting all random patterns of up to $\frac{n-k}{\log_2(n-1)}$ number of errors, there are several decoding algorithms for BCH codes proposed in the literature. Among these decoding algorithms, here, ordered statistics (OS) decoder is the one which attracts more attention compared to the others, due to its performance for short block-lengths and parametrized structure, that can be changed with a single variable, i.e. the order of the OS decoder. Although it is proven that there does not exist any BCH code that can achieve the capacity as $n \to \infty$, but it is shown in [61] and [39] that the performance of an extended BCH code with OS decoder performs better than the other candidate codes and approach to the normal approximation of the maximum achievable rate for short block-lengths.

- **Convolutional Codes with Viterbi Decoder**: Convolutional codes are one of the powerful and easy to implement codes. Today, they are used in several standards including IEEE 802.11 and satellite communication standards as well. Convolutional encoding contains memories and has a different encoding structure than the linear block codes. Due to the memories, outputs at any given time unit depend on the inputs of the instantaneous time and on the m previous inputs, where m is the number of memories in the encoding scheme.

Convolutional codes are first presented by Elias in 1955 [17] as an alternative to linear block code. An efficient decoding algorithm, which is based on finding the most likely codeword from among the set of all possible codewords, is presented by Viterbi in 1967 [62]. The Viterbi algorithm excessively decreases the decoding complexity, but the complexity still scales exponentially with the number of memories. It is numerically shown in [21] that, compared to other coding schemes, convolutional codes with Viterbi decoder perform good and approach the maximum achievable bound. However, this happens with high memory orders, which reveals a trade-off for complexity constrained receivers.

- **LDPC Codes with Belief Propagation Decoding**: Low Density Parity Check (LDPC) codes have been first proposed by Gallager [20] in 1962. But due to its computational burden they could not be used practically until 1997 when they were rediscovered by Mackay [41]. Today, thanks to the developments made in the field of integrated circuits, LDPC codes are being used in several wireless standards such as IEEE 802.11n.

  LDPC codes are linear codes and have low density, i.e. having relatively small number of 1s in the parity check matrix, the decoding process is computationally efficient. However, although LDPC codes with the iterative bilief propagation decoding approach capacity with long codewords, e.g. $n > 10^5$, their performance degrades for moderate or short codeword lengths [39].
- **Polar Codes with Successive Cancellation Decoder**: Polar codes are linear codeword error correcting codes and first discovered by Arikan in 2009 [6]. It is proven that polar codes with successive cancellation decoder are capable of achieving the capacity for binary memoryless symmetric channels when the code length approaches infinity. Therefore, polar codes are proposed to be used in 5G communications systems [59]. However, for short length codes, successive cancellation decoding fail to satisfy reasonable error-correction performances compared to other coding schemes. This issue has been solved by list decoding, albeit the expense of computational complexity [59].

Performance comparisons among these channel coding schemes are presented in Figs. 2.7 and 2.8 for several (128,64) and (128,85) codes over the Bi-AWGN channel.[2] The finite-length performance benchmark is the normal approximation shown in (2.11) and the benchmark error rate is computed as

$$\epsilon_m = Q\left(\sqrt{\frac{n}{\mathcal{V}(\rho)}}\frac{(C(\rho) - r)}{\log_2 e}\right). \tag{2.17}$$

[2]CER results for convolutional codes, LDPC codes, and polar codes are taken from [38].

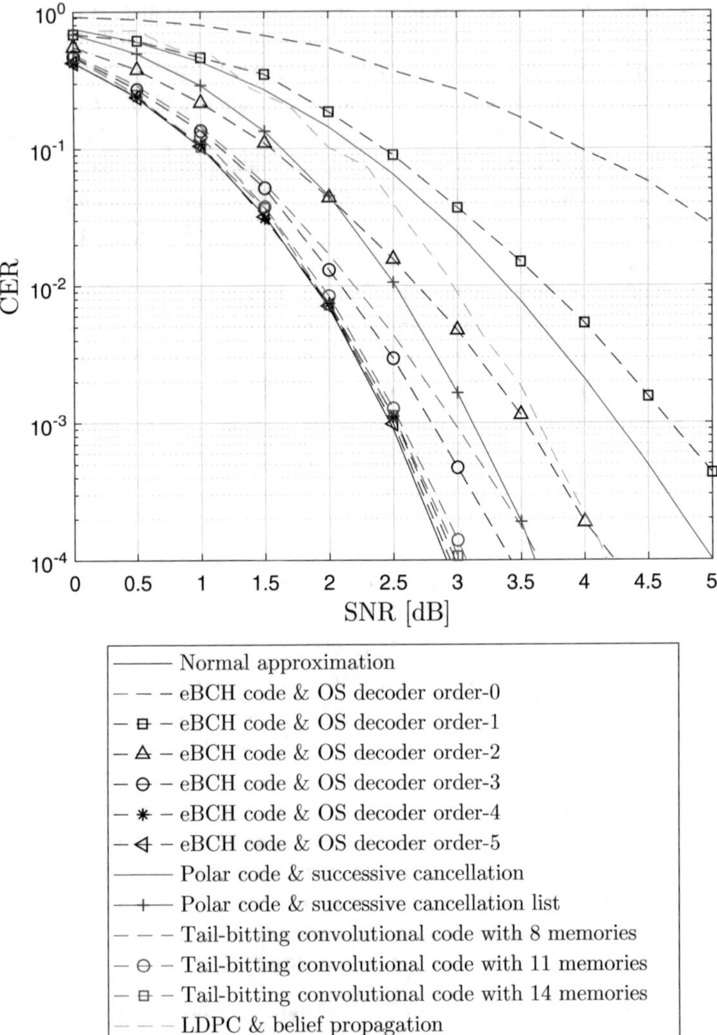

Fig. 2.7 CER for several (128,64) codes over the Bi-AWGN channel

Performance of OS decoders with several orders show that it approaches the benchmark as the order becomes sufficiently high.

Similar to the OS decoder, tail-biting convolutional codes are also performs better than most of the codes and their performance approach to the normal approximation as the memory number increases. On the other hand, performance of LDPC codes

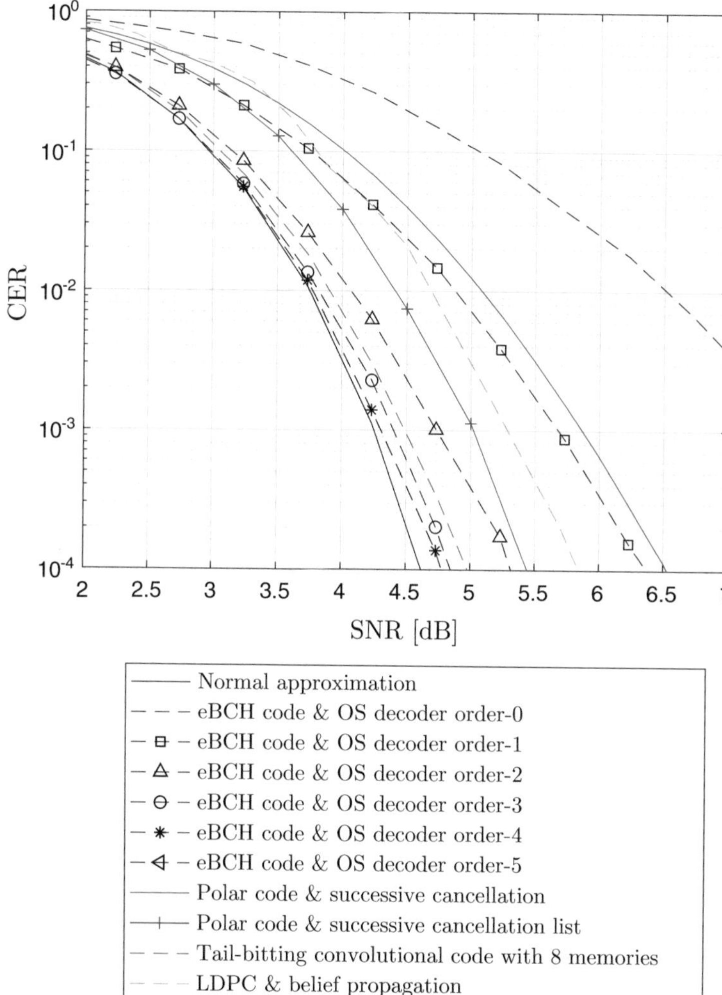

Fig. 2.8 CER for several (128,85) codes over the Bi-AWGN channel

with belief propagation decoder and polar codes with successive cancellation and successive list decoding are also depicted in the figure. Although, LDPC codes have slightly better performance than the polar codes, both codes' performance fall behind compared to the others.

For the analysis of low-latency communication for low-complexity IIoT receivers, we focus on BCH codes with OS decoders, since empirical results show that they perform better relative to other short block length codes.

2.4 Low-Latency Communication for Low-Complexity IIoT Receivers

Both in asymptotic and non-asymptotic information theoretic studies, it is mostly assumed that the decoding duration is negligible. This assumption is reasonable for loose latency constraints, small number of information bits and channel uses or unlimited computation power at the receiver. However, these assumptions cannot be justified for complexity constrained receivers, such as mission-critical IIoT devices, due to the limited processor capabilities. In this case, the decoding time need to be calculated carefully and included in the total latency analysis.

For the rest of the chapter, we focus on low-latency communication for complexity constrained receivers. Here, based on a comprehensive model, that is presented in [10], interesting trade-offs among decoding complexity (in number of binary operations per information bit), code rate, codeword length, CER, and power budget will be presented. In particular, we first consider the complexity of the maximum-likelihood (ML) and OS decoders. Then, we will show their computational complexities in terms of number of binary operations per information bit and a simple but consistent way to calculate the decoding time of a complexity constrained receiver is presented. Using the model from [10], one can easily calculate the power penalty from the normal approximation that is needed to ensure the reliability. Finally, some communication scenarios which shows the significance of decoding duration in latency constrained communication are presented.

2.4.1 System Model

A communication medium described in Fig. 2.3 in which communication over a Bi-AWGN channel is considered. Thus, $u_i \in \{0, 1\}$, $x_i \in \{-\sqrt{\rho}, +\sqrt{\rho}\}$, the received sample $y_i \in \mathbb{R}$, and h is the deterministic channel coefficient that is known to the receiver and transmitter. Suppose that the encoder at the transmitter is a linear block code encoder where the channel inputs are obtained as follows

$$\boldsymbol{x} = \boldsymbol{u} \odot \boldsymbol{G} \tag{2.18}$$

where $\odot$ denotes Boolean multiplication and $\boldsymbol{G}$ is the generator matrix in systematic form such as

$$\boldsymbol{G} = [\boldsymbol{I}|\boldsymbol{P}], \tag{2.19}$$

where $\boldsymbol{I}$ and $\boldsymbol{P}$ are the $k \times k$ identity matrix and $k \times (n-k)$ parity check matrix, respectively. Thus, the channel encoder takes k number of information bits and produces an n length of codeword and the transmission rate is $r = \frac{k}{n}$. The noisy channel output $\boldsymbol{y}$ is received at the receiver and the decoder maps the received sequence into an estimate of the message.

The normal approximation that is presented in Sect. 2.3.3 is used as the benchmark for the maximum achievable information rate over the Bi-AWGN channel and $C(\rho)$ and $\mathcal{V}(\rho)$ for Bi-AWGN channel are expressed as

$$C(\rho) = \max_{p_X(x)} I(X;Y) \tag{2.20}$$

$$= \frac{1}{2\pi} \int_{-\infty}^{\infty} e^{-\frac{z^2}{2}} \left(1 - \log_2\left(1 + e^{-2\rho + 2z\sqrt{\rho}}\right)\right) \mathrm{d}z\,, \tag{2.21}$$

$$\mathcal{V}(\rho) = \mathrm{var}\left[\imath(X;Y)\right] \tag{2.22}$$

$$= \frac{1}{2\pi} \int_{-\infty}^{\infty} e^{-\frac{z^2}{2}} \left(1 - \log_2\left(1 + e^{-2\rho + 2z\sqrt{\rho}}\right) - C(\rho)\right)^2 \mathrm{d}z\,, \tag{2.23}$$

where $\mathrm{var}\left[\imath(X;Y)\right]$ denotes the variance of mutual information density $\imath(X;Y)$ under the capacity achieving input distribution $p_x(x)$.

The total communication latency arising from transmission and decoding durations for a codeword of n symbols can be written as

$$d_t = nT_s + d_D, \tag{2.24}$$

where T_s is the time duration of a single channel use and d_D is the time required for the decoding. A detailed information about the ML and OS decoders are given in the following sections.

2.4.2 *Decoding Complexity*

2.4.2.1 ML Decoder

The optimal decoder that minimizes the probability of error for equiprobable codewords is the ML decoder. It calculates and compares the Euclidean distances between the observed vector and all possible codewords and selects the one that minimizes the distance. This process can be analytically shown as

$$\hat{\boldsymbol{x}}_{\mathrm{ML}} = \arg\min_{\boldsymbol{x}\in\mathcal{X}} \|\boldsymbol{y} - h\boldsymbol{x}\| \tag{2.25}$$

$$= \arg\max_{\boldsymbol{x}\in\mathcal{X}} \sum_{i=1}^{n} \mathrm{Re}\{y_i h^* x_i^*\} \tag{2.26}$$

where $\mathcal{X}$ represents the set of all possible codewords. The number of operations per information bit, in terms of codeword comparisons $c_{\mathrm{ML}} = 2^{k-1}$. Thus, the computational complexity of the optimal decoder is exponential in k. This complexity value can also be accepted as an upper bound on the complexity of any practical decoder. However, since c_{ML} becomes very high even for small number of information bits, ML decoder is infeasible for real-world applications. Therefore, less complex decoders with the mission to approach to the performance of ML decoder are proposed in the literature such as OS decoders for linear block codes [19], Viterbi decoder for convolutional codes [62].

2.4.2.2 OS Decoder

For a linear block code (n, k, d_{min}), where d_{min} represents the minimum Hamming distance between any two codewords, a brute-force approach to ML decoding is generally impossible for non-trivial codes, since the computational demand is too high.

Various different decoding algorithms have been proposed in the literature for decoding of linear block codes. In this chapter we will focus on the OS decoder which is a universal near-optimal soft decoder that can decode any linear block code. OS decoder is based on ordered statistics of the received noisy codeword and the decoding process can be briefly summarized as follows:

- Sort the received vector, $\boldsymbol{y}$, with respect to the absolute amplitude values and form the sorted binary vector $\boldsymbol{y}'$,

$$\boldsymbol{y}' = \kappa(\boldsymbol{y}), \text{ where } |y_1'| \geq |y_2'| \geq \cdots \geq |y_n'|. \tag{2.27}$$

  where $\kappa(\cdot)$ is the permutation pattern.
- Hard-decode $\boldsymbol{y}'$ to produce

$$\boldsymbol{y}^* = \text{sign}(\boldsymbol{y}'), \quad \text{where } \text{sign}(z) = \begin{cases} 0, & \text{if } z < 0, \\ 1, & \text{if } z \geq 0. \end{cases} \tag{2.28}$$

- Re-order the columns of the generator matrix such that $\boldsymbol{G}' = \kappa(\boldsymbol{G})$ and apply the Gauss-Jordan elimination to form the new systematic generator matrix $\boldsymbol{G}^*$.
  - Note that it is possible that the first k columns of $\boldsymbol{G}'$ may be linearly dependent. In this case, reaching to a new systematic generator matrix $\boldsymbol{G}^*$ is not possible. Therefore, a second permutation is needed in order to have k independent columns.
- Generate the set of error vectors, denoted as $\mathcal{T}$, depending on the order$-s$.
- Encode $\boldsymbol{y}^* \oplus \boldsymbol{t}_i$ with $\boldsymbol{G}^*$, where $\boldsymbol{t}_i \in \mathcal{T}$ is the ith error vector and $\oplus$ represents the binary addition.
- Search over $\boldsymbol{t}_i$ that minimizes the Euclidean distance between the encoded binary vector and $\boldsymbol{y}'$ (see Fig. 2.9).

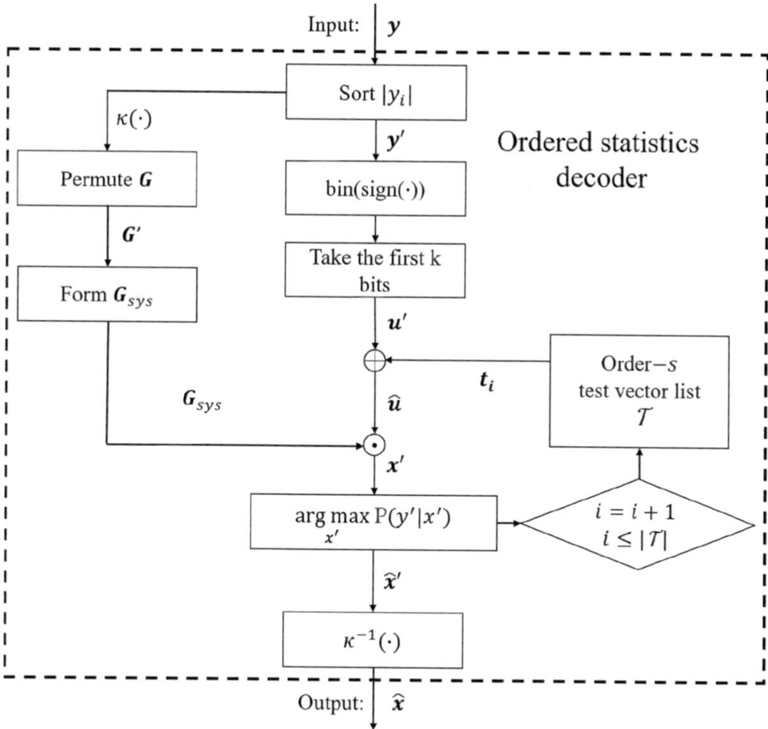

Fig. 2.9 Block diagram of an OS decoder

Hence, the OS decoding of $\boldsymbol{y}$ can be formulated as

$$\hat{\boldsymbol{x}}_{OS} = \kappa^{-1}\left(\underset{\{\hat{\boldsymbol{x}}|\ \boldsymbol{t}\in\mathcal{T},\ \hat{\boldsymbol{x}}=(\boldsymbol{y}^*\oplus\boldsymbol{t})\odot\boldsymbol{G}^*\}}{\operatorname{argmax}} \mathbb{P}(\hat{\boldsymbol{x}}|\boldsymbol{y}')\right) \tag{2.29}$$

$$= \kappa^{-1}\left(\underset{\{\hat{\boldsymbol{x}}|\ \boldsymbol{t}\in\mathcal{T},\ \hat{\boldsymbol{x}}=(\boldsymbol{y}^*\oplus\boldsymbol{t})\odot\boldsymbol{G}^*\}}{\operatorname{argmin}} \|\boldsymbol{y}' - \hat{\boldsymbol{x}}\|^2\right), \tag{2.30}$$

where $\kappa^{-1}(\cdot)$ applies the inverse permutation to the codeword that minimizes the Euclidean distance. Note that, the key item of the decoder is the order$-s$ of the decoder since it determines the range of the codeword search and substantially affect the probability of erroneous estimates. The cardinality of $\mathcal{T}$ for order$-s$ is

$$|\mathcal{T}| = \sum_{i=0}^{s} \binom{k}{i}. \tag{2.31}$$

In fact, this number is the total number of codeword comparisons and polynomially increases in the order of k^s. As $s \rightarrow k$ the cardinality of $\mathcal{T}$ approaches 2^k since $\sum_{i=0}^{k}\binom{k}{i} = 2^k$ and hence the decoding process will consider all possible codewords and the performance will be equivalent to the ML decoding. However, in order to approach the ML decoder performance, it is shown in [19] that assuming an AWGN channel, the recommended decoding order is

$$s_r = \min\left\{\left\lceil \frac{d_{\min}}{4} - 1\right\rceil, k\right\}, \tag{2.32}$$

where $d_{\min}$ represents the minimum Hamming distance of the codeword set and $\lceil \cdot \rceil$ is the ceiling function. Therefore, at least

$$2^k - \sum_{i=0}^{s_r}\binom{k}{i} \tag{2.33}$$

number of unnecessary codeword comparisons are saved.

Performances of OS decoder with different orders where $k = 64$, $n = 128$, and $d_{\min} = 22$ are shown in Fig. 2.10. The information bits are encoded with an BCH encoder and ϵ_m represents the normal approximation error rate bound derived from (2.11). Figure 2.10 shows that as the order$-s$ increases, the performance of the

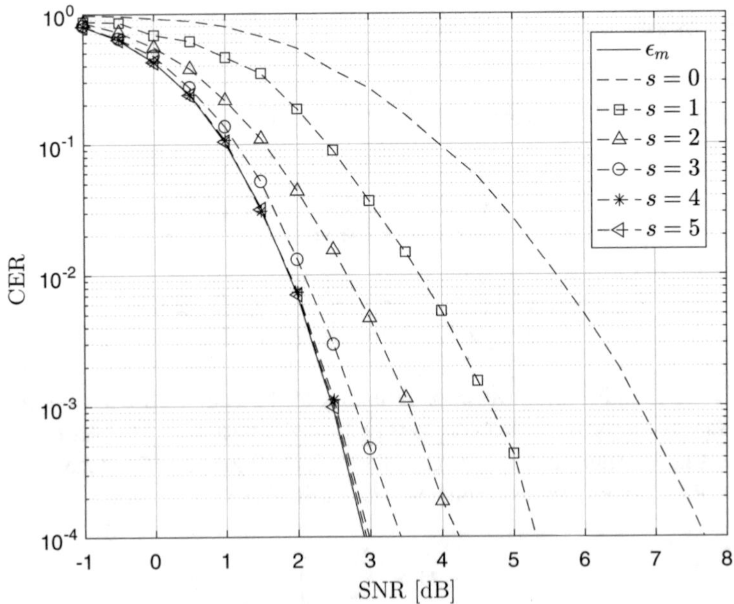

Fig. 2.10 CER performace of OS decoder with different orders compared to the normal approximation (2.17) error rate bound for Bi-AWGN channel where $n = 128$ and $k = 64$

OS decoder increases and reaches ϵ_m when $s_r = 5$. Note that although increasing the order$-s$ boosts up the performance, it also leads to an increase in the computational complexity of the decoder as it will be shown in the next Section.

2.4.2.3 Computational Complexity and Decoding Duration

A general expression that shows the computational complexity of all codes is unlikely to be found. Therefore, in this section, we study the BCH codes with OS decoders and propose a system model based on them. As mentioned, the decoding complexity of the ML decoder in terms of codeword comparisons per information bit is

$$c_{\mathrm{ML}} = 2^{k-1}. \tag{2.34}$$

On the other hand, it is shown in [10] that for OS decoders the total number of binary operations per information bit can be calculated as

$$c_{\mathrm{OS}} = \frac{k^2}{8} + \frac{n}{2}\sum_{i=0}^{s}\binom{k}{i}. \tag{2.35}$$

Therefore, the complexity order of the OS decoder in terms of binary operations per information bit can be written as

$$c_{\mathrm{OS}} = \begin{cases} O(k^2), & \text{if} s \leq 2, \\ O(k^s), & \text{if} s > 2, \end{cases} \tag{2.36}$$

since as $s \leq 2$ complexity is dominated by the Gauss-Jordan elimination process of the permuted matrix $\boldsymbol{G}'$. However, for $s > 2$, since the number of codeword comparisons rises polynomially, the second part dominates the complexity.

The decoding duration is directly related with the particular hardware. Although, computing the total execution time of a software program after having the total number of binary operations is hard, without loss of generality, one can relate the execution time by estimating the required amount of clock cycles of a processor. Therefore, for simplicity and generality, a linear relation between d_D and the time required for a binary operation on the hardware platform that the decoder operates, denoted as T_b, is assumed. Then the total latency for the transmission of a codeword of length n is

$$\begin{aligned} d_t &= nT_s + d_D \\ &= nT_s + kcT_b. \end{aligned} \tag{2.37}$$

Accuracy of d_t can be improved further by evaluating the hardware technology, architectural design, memory timings, etc. But since these are beyond the scope of this chapter, we confine to (2.37) for further analysis.

A constraint on total latency requests that the information shall be encoded, transmitted, received and decoded at the receiver within the total allowed time duration. This is especially significant for mission-critical communication, where any additional latency on data transmission may cause severe affects. Therefore, one can assume that if a codeword cannot be transmitted within the time constraint is counted as a fail.

Suppose that the total time constraint on d_t is defined as

$$d_t = nT_s + kcT_b \leq d_m, \tag{2.38}$$

where d_m represents the maximum latency deadline. Such a case can be visualised as in Fig. 2.11. Note that for fized n, d_m restricts the order$-s$ as follows

$$s \leq s_m = \underset{\{s|s\in\mathbb{Q}^+,\ nT_s+ckT_b\leq d_m\}}{\arg\max} \quad c, \tag{2.39}$$

Transmission and decoding latencies for a particular receiver:

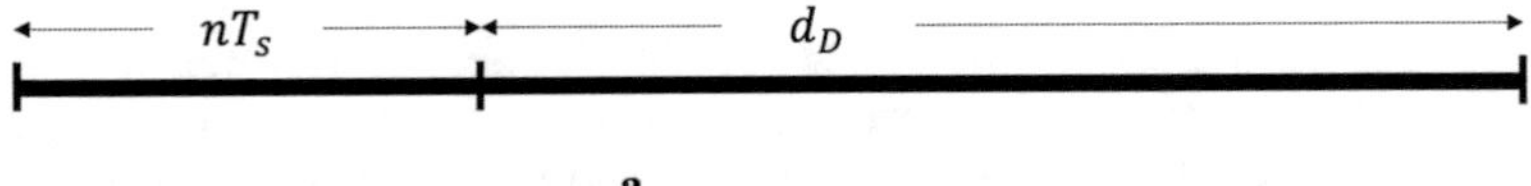

a

Constraint on $nT_s + d_D$ is applied such as : $nT_s + d_D \leq d_m$

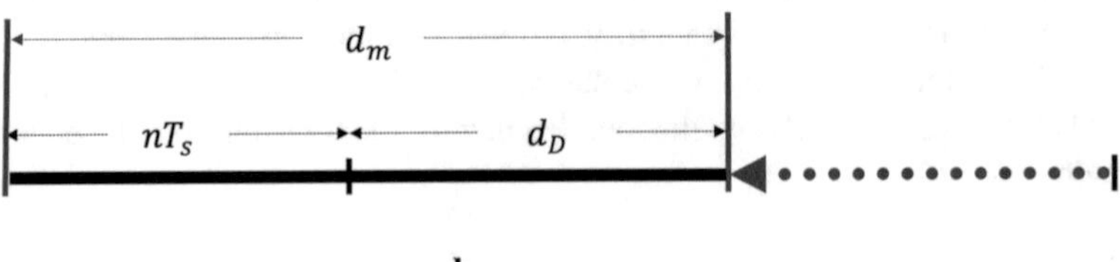

b

Fig. 2.11 A possible circumstance when a constraint shown in (2.38) is imposed

where s_m denotes the maximum allowed order. Although it is hard to derive a closed form expression for s_m, an approximation has been derived in [10]

$$s_m \approx \frac{k}{2}\left(1 - \sqrt{1 - \left(\frac{\log_2 \tau}{k}\right)^{4/3}}\right), \tag{2.40}$$

where

$$\tau = \frac{2(d_m - nT_s)}{nkT_b} - \frac{k^2}{4n}. \tag{2.41}$$

However, a constraint on order$-s$ may lead to a degradation in the performance of the decoder and if

$$s_m < s_r \tag{2.42}$$

the performance may not even approach to the ML decoder performance. This problem raises interesting paradigms in between the variables such as n, k, γ, ϵ, and c, which will be presented in the following Section.

2.4.3 Constraints on Computational Complexity vs Power Penalty

Let the transmission and decoding latencies for a particular receiver are as depicted in Fig. 2.11a. Suppose a latency constraint, introduced in (2.38), is imposed and therefore the communication system is faced with a circumstance that is shown in Fig. 2.11b. In order to meet the latency constraint, the total latency d_t of the communication systems can be controlled by the selection of order$-s$, as bounded in (2.39), for a particular code of fixed n and k. However, in this case the cost would be the reduced reliability, i.e. higher ϵ. A plausible way to reduce ϵ and achieve the desired reliability while meeting the latency constraint is to increase the signal power. This increment in power is called the power penalty resulting from the latency constraint. Hence, one can see the complex relation between the total latency, power penalty, and complexity of the decoder.

As the order$-s$ of the decoder increases, computational complexity and performance of the decoder increases. Therefore, a direct proportion between c_{OS} and ϵ is expected. Results shown in [10] and [56] reveal that as the performance of an OS decoder approaches to the normal approximation, its complexity exponentially increases. It is also shown in [10] that the relation between power penalty and computational complexity can be closely approximated by

$$\log_2 c = \frac{1}{a(\Delta\rho)^{\gamma} + b}, \tag{2.43}$$

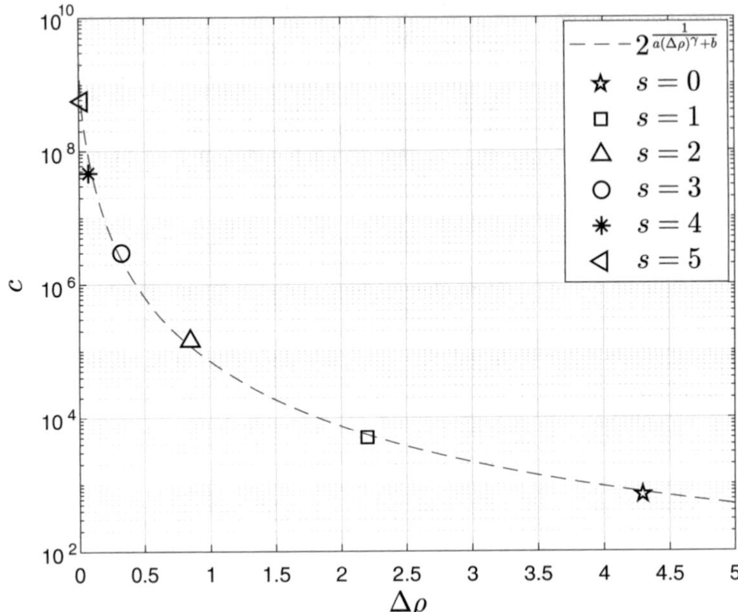

Fig. 2.12 Power penalty values of OS decoder at different orders versus their complexity for $n = 128$, $k = 64$, and $\epsilon = 10^{-3}$

for proper selections of the positive constants a, γ and b. $\Delta\rho$ represents the power penalty from the normal approximation for the maximum achievable coding rate. Note that $\Delta\rho \geq 0$, since no coding rate is achievable above the maximal coding rate given by (2.11).

An illustration of this behavior is depicted in Fig. 2.12 for $n = 128$, $k = 64$, and $\epsilon = 10^{-3}$, where the vertical and horizontal axis represent the total number of binary operations per information bit and the power penalty, respectively. Note that the dashed line is the model proposed in (2.43) and each marker represents an order$-s$ of the OS decoder and shows the power penalty of an order$-s$ decoder at a fixed rate versus its complexity. Note that these power values are computed by searching the required amount of SNR by gradually increasing it until CER reaches ϵ.

Figure 2.12 shows that the computational complexity of the decoder increases exponentially as the power penalty goes to zero, i.e. the decoder approaches the normal approximation.

Hence, for a fixed reliability constraint, (2.43) reveals the trade-off between computational complexity and power penalty in a simple and tractable way. Although (2.43) is based on BCH codes with OS decoders, similar relations are observed in [56] for different coding schemes such as polar codes and tail-biting convolutional codes. Hence, one can claim that (2.43) is crucial for the analyses of URLLC with computational complexity constraints.

2.4.4 Numerical Examples

In this section, we investigate different communication scenarios with latency, reliability and computational complexity constraints.

2.4.4.1 Maximal Information Rate

The theoretical background on maximal information rate for finite blocklength is presented in Sect. 2.3.3. However, as mentioned, these analyses neither include the decoding time nor any constraint on the computation complexity of the decoder.

The maximal information rate versus the SNR for $n = 128$ and $\epsilon = 10^{-3}$ is shown in Fig. 2.13 where ergodic capacity and normal approximation are represented by the dashed and solid lines, respectively. The dotted line denotes the maximal information rate when a total latency constraint of $d_m = 1$ ms is imposed

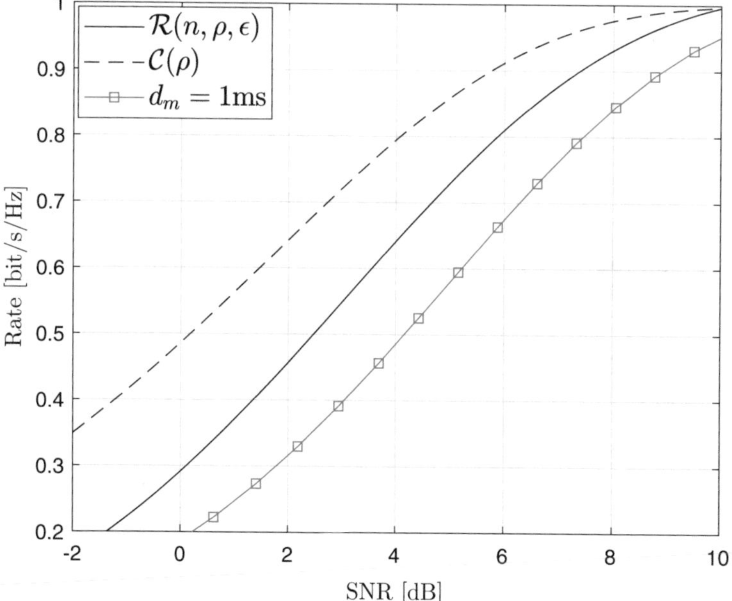

Fig. 2.13 New achievability bounds under latency and complexity constraints for $n = 128$, $\epsilon = 10^{-3}$, $T_s = 1\,\mu$s, and $T_b = 1$ ns

to the communication system, where $T_s = 1\,\mu$s and $T_b = 1$ ns. Note that these values can be related to a communication system with transmission rate at 1 MHz and a receiver with 1 GHz processor speed.

In order to create Fig. 2.13, the maximum allowable decoding time is calculated using (2.37) for each rate and n. By doing so, the amount of extra power, namely the power penalty, is calculated by (2.43). Hence, the dotted curve is achieved by shifting normal approximation by $\Delta\rho$ to right, since $\Delta\rho \geq 0$. Thus, one can conclude that the latency constraint causes a back off from the maximal achievable rate or a power penalty need to be added.

2.4.4.2 Maximization of *k*

Suppose an IIoT communication scenario where the transmitter with a total power budget of ρ_m intends to transmit a number of information bits subject to a total transmission and decoding latency constraint of d_m with codeword error probability ϵ. In order to maximize the communication efficiency, the transmitter needs to maximize the number of information bits k that can be transmitted with the total power budget within d_m duration.

Let us first investigate the unlimited computation power case. A received noisy codeword can be decoded suddenly in this case, since $T_b = 0$ ns. Thus, all the allowed latency duration can be used for transmission of the codeword such that

$$n_m = \left\lfloor \frac{d_m}{T_s} \right\rfloor \tag{2.44}$$

number of symbols can be transmitted at the highest possible rate that is allowed by the power budget. This limit is determined by (2.11), which yields

$$k_m = \lfloor n_m \mathcal{R}(n_m, \epsilon) \rfloor , \tag{2.45}$$

where k_m represents the maximized k.

However, selection of k_m is not trivial for a computational complexity constrained receiver since interesting trade-offs arise. Suppose that n is selected small, in this case, the total duration that is allocated for decoding is long enough that may allow to use a high code rate. However, as n increases, the total duration for decoding reduces and therefore a coding scheme with lower code rate need to be selected in order to meet the latency constraint.

Maximum number of k for each realization of n is depicted in Fig. 2.14 for $d_m = 1$ ms, $\rho_m = 3$ dB, and $\epsilon = 10^{-3}$. To see the effect of computation constraints three different choices of T_b are selected such as $T_b = \{10, 1, 0\}$ ns, where $T_b = 0$ ns represent infinite computation power and $T_b = \{10, 1\}$ ns corresponds to 100 MHz and 1 GHz of processor computation power, respectively.

The trade-off that is explained previously can be seen in Fig. 2.14 clearly. Note that for each case, there is a maximum that appears at different selections of n

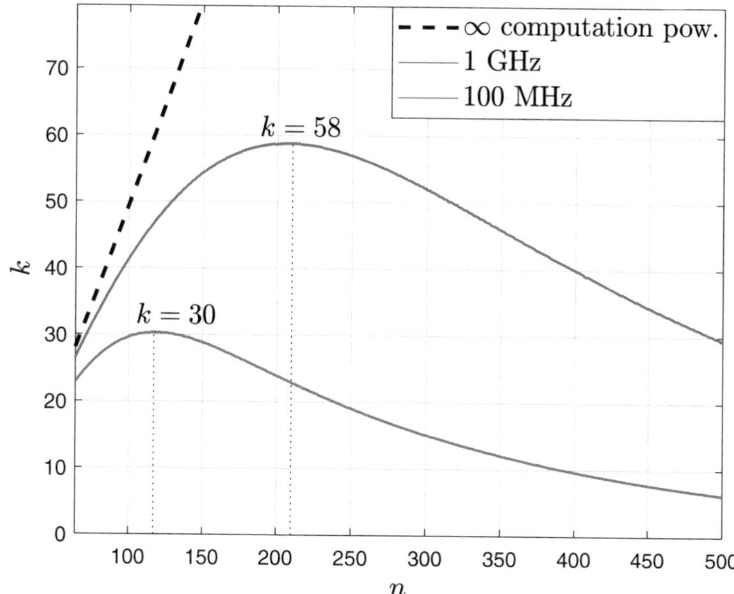

Fig. 2.14 Maximization of k where $d_m = 1$ ms, $\rho_m = 3$ dB, and $\epsilon = 10^{-3}$

and k. Optimum n values are found at $n = \{117, 210, 1000\}$ and corresponding k_m values are $k_m = \{30, 58, 649\}$. Ratios between k_m for complexity constrained receivers to (2.45) are 0.046, 0.089. Hence, it is seen that if complexity constraints and decoding duration are taken into account, depending on the receiver capabilities, the maximum number of k that can be transmitted is far less than the theoretical limits.

2.4.4.3 Minimization of d_t

Suppose that a fixed number of information bits, k, subject to a codeword error probability constraint, ϵ and a maximum power constraint, ρ_m, is intended to be transmitted. But the optimum selection of n and r is searched that minimizes to total latency, d_t. This question may arise in the perspective of the system designer when a fixed number of information bits is intended to be transmitted subject to a power budget.

For fixed k, selecting a small n that corresponds a higher code rate leads to very large codebook. Therefore, the total duration of the decoding increases exponentially. In this case, two cases may occur; the required coderate may exceed (2.11) and therefore the transmission is not possible or the selected code rate may be very close to the normal approximation, which leads to a very high computational complexity which yields to very high decoding duration. This can

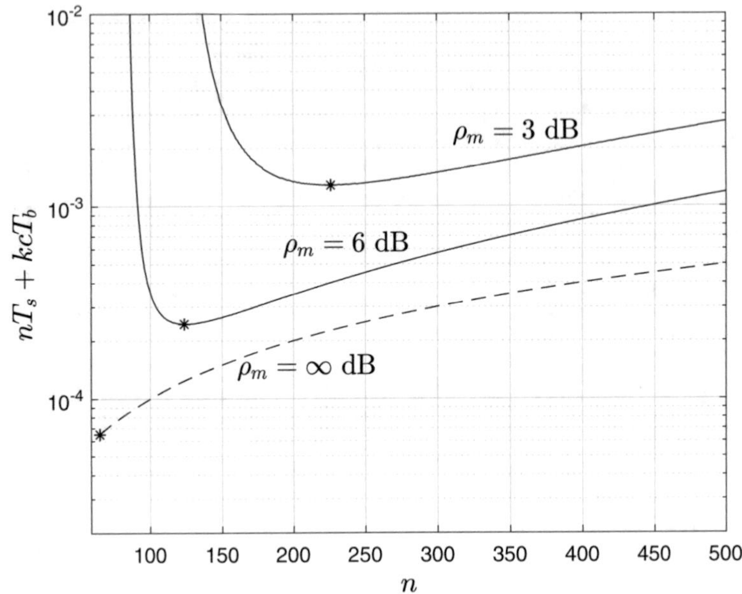

Fig. 2.15 Minimum d_t with respect to n for several SNR where $k = 64$, $\epsilon = 10^{-3}$, and $T_b = 10^{-9}$ s

be decreased by increasing n. As n increases, the code rate and the codebook size decreases as well. This will decrease the complexity of the decoder since the code rate may be sufficiently far from the normal approximation. In this case, it is more likely to find a less complex decoder that can be supported by the power budget.

In Fig. 2.15 the minimum total latency for each n is plotted. The optimum n can be selected as the one that minimizes all. It can also be seen in Fig. 2.15 that an optimum point d_t increases. This happens when the code rate is so low that the duration d_D is negligible compared to nT_s and therefore the total latency is dominated by the codeword transmission latency. In Fig. 2.15 the optimal selection of n values are $n_{\text{opt}} = \{226, 124\}$ for power constraints $\rho_m = \{3, 6\}$ dB, respectively.

In order to compare the results with a benchmark, the noiseless case where $P_m = \infty$ dB is also plotted in the Figure. In this case, according to (2.43), since there is no error in the received codeword, $d_D = kT_b \approx 0$ s and $d_t = nT_s$ and d_t linearly increases in n. Thus, the optimum selection of n is $n_{\text{opt}} = k$.

2.5 Conclusions

In this chapter, the focus is on wireless communication for IIoT. Initially, the state-of-the-art wireless standards and technologies, which are suitable for IIoT applications, are presented. Further, the next generation of mobile wireless communications, namely 5G, and its three major usage scenarios are introduced. URLLC, which is one of these scenarios and crucial for mission-critical IIoT applications, is investigated deeply. Challenges of enabling URLLC for low complexity IIoT devices are studied. Therefore, the theoretical limits on the information transmission of wireless channels are presented and based on a proper system model, interesting trade-offs are revealed for complexity constrained IIoT nodes when a constraint on transmission and decoding latency is applied. Results reveal that for receivers with computational complexity receivers decoding time has a considerable effect on the bounds of the short blocklength codes when there is a latency constraint on the system. It is shown that in these cases the optimal selection of operating points become a non-trivial task which gives rise to a number of interesting optimization problems.

Acknowledgments This work was funded in part by the Swedish Foundation for Strategic Research (SSF) under grant agreement RIT15-0091.

References

1. Afaqui, M.S., Garcia-Villegas, E., Lopez-Aguilera, E.: IEEE 802.11ax: challenges and requirements for future high efficiency WiFi. IEEE Wireless Commun. **24**(3), 130–137 (2017)
2. Akpakwu, G.A., Silva, B.J., Hancke, G.P., Abu-Mahfouz, A.M.: A survey on 5G networks for the internet of things: communication technologies and challenges. IEEE Access **6**, 3619–3647 (2018)
3. Alohali, B.A., Vassilakis, B.A., Moscholios, I.D., Logothetis, M.D.: A secure scheme for group communication of wireless IoT devices. In: 11th International Symposium on Communication Systems. Networks & Digital Signal Processing, Budapest (2018) pp. 1–6
4. Anand, A., De Veciana G., Shakkottai, S.: Joint scheduling of URLLC and eMBB traffic in 5G wireless networks. In: IEEE INFOCOM - IEEE Conference on Computer Communications, Honolulu (2018), pp. 1970–1978
5. Apsel, A.: A simple guide to low-power wireless technologies: balancing the tradeoffs for the internet of things and medical applications. IEEE Solid-State Circuits Mag. **10**(4), 16–23 (2018)
6. Arikan, E.: Channel polarization: a method for constructing capacity-achieving codes for symmetric binary-input memoryless channels. IEEE Trans. Inf. Theory **55**, 3051–73 (2009)
7. Bennis, M., Debbah, M., Poor, H.V.: Ultrareliable and low-latency wireless communication: tail, risk, and scale. Proc. IEEE **106**(10), 1834–1853 (2018)
8. Bose, R.C., Ray-Chaudhuri, D.K.: On a class of error correcting binary group codes. Inf. Control **3**(3), 279–290 (1960)
9. Celebi, H.B., Pitarokoilis, A., Skoglund, M.: Training-assisted channel estimation for low-complexity squared-envelope receivers. In: IEEE 19th International Workshop on Signal Processing Advances in Wireless Communications (SPAWC), Kalamata (2018)
10. Celebi, H.B., Pitarokoilis, A., Skoglund, M.: Low-latency communication with computational complexity constraints. In: International Symposium on Wireless Communication Systems (2019)
11. Chakrapani, A.: Efficient resource scheduling for eMTC/NB-IoT communications in LTE Rel. 13. In: IEEE Conference on Standards for Communications and Networking (CSCN), Helsinki (2017)

12. Chang, K.: Interoperable nan standards: a path to cost-effective smart grid solutions. IEEE Wireless Commun. **20**(3), 4–5 (2013)
13. Chen, B., Wan, J., Shu, L., Li, P., Mukherjee, M., Yin, B.: Smart factory of industry 4.0: key technologies, application case, and challenges. IEEE Access **6**, 6505–6519 (2018)
14. Costa-Requena, J., Poutanen, A., Vural, S., Kamel, G., Clark, C, Roy, S.K.: SDN-based UPF for mobile backhaul network slicing. In: European Conference on Networks and Communications (EuCNC), Ljubljana, Slovenia (2018), pp. 48–53
15. Destounis, A., Paschos, G.S., Arnau, J., Kountouris, M.: Scheduling URLLC users with reliable latency guarantees. In: 16th International Symposium on Modeling and Optimization in Mobile, Ad Hoc, and Wireless Networks (WiOpt), Shanghai (2018), pp. 1–8
16. Domazetovic, B., Kocan, E., Mihovska, A.: Performance evaluation of IEEE 802.11ah systems. In: 2016 24th Telecommunications Forum (TELFOR), Belgrade (2016) pp. 1–4
17. Elias, P.: Coding for Noisy Channels. IRE Conv. Rec., Part 4 (1955), pp. 37–47
18. ETSI: Electromagnetic compatibility and Radio spectrum Matters; System Reference Document; Short Range Devices; Part 2: Technical characteristics for SRD equipment for wireless industrial applications using technologies different from Ultra-Wide Band (2011)
19. Fossorier, M.P.C., Lin, S.: Soft-decision decoding of linear block codes based on ordered statistics. IEEE Trans. Inf. Theory **41**(5), 1379–1396 (1995)
20. Gallager, R.: Low-density parity-check codes. IRE Trans. Inf. Theory **8**(1), 21–28 (1962)
21. Gaudio, L., Ninacs, T., Jerkovits, T., Liva, G.: On the performance of short tail-biting convolutional codes for ultra-reliable communications. In: 11th International ITG Conference on Systems, Communications and Coding, Hamburg (2017)
22. GSMA Intelligence: The Mobile Economy, GSMA Intelligence Report (2019). Available at https://www.gsma.com/mobileeconomy/
23. Goldsmith, A.: Wireless Communications. Cambridge University Press, Cambridge (2005)
24. Hassan, S.M., Ibrahim, R., Bingi, K., Chung, T.D., Saad, N.: Application of wireless technology for control: a wireless HART perspective. Proc. Comput. Sci. **105**, 240–247 (2017)
25. Hocquenghem, A.: Codes correcteurs d'erreurs. Chiffres **2**, 147–156 (1959)
26. Hui, J.W., Culler, D.E.: Extending IP to low-power, wireless personal area networks. IEEE Internet Comput. **12**(4), 37–45 (2008)
27. IEEE 802.11ax: The sixth generation of Wi-Fi. Cisco Public Technical White Paper (2018)
28. IEEE Standard for Information technology-Telecommunications and information exchange between systems Local and metropolitan area networks—Specific requirements - Part 11: Wireless LAN Medium Access Control and Physical Layer Specifications: in IEEE Std 802.11-2016, 14 Dec (2016)
29. IMT Vision-Framework and Overall Objectives of the Future Development of IMT for 2020 and Beyond, document Recommendation ITU-R M.2083-0 (2015). Available at https://www.itu.int/dms_pubrec/itur/rec/m/R-REC-M.2083-0-201509-I!!PDF-E.pdf
30. Jewel, M.K.H., Zakariyya, R.S., Famoriji, O.J., Ali, M.S., Lin, F.: A low complexity channel estimation technique for NB-IoT downlink system. In: IEEE MTT-S International Wireless Symposium (IWS), Guangzhou (2019)
31. Kadambar, S., Reddy Chavva, A.K.: Low complexity ML synchronization for 3GPP NB-Io. In: International Conference on Signal Processing and Communications (SPCOM), Bangalore (2018)
32. Karimi, A., Pedersen, K.I., Mahmood, N.H., Steiner, J., Mogensen, P.: 5G centralized multi-cell scheduling for URLLC: algorithms and system-level performance. IEEE Access **6**, 72253–72262 (2018)
33. Lekomtcev, D., Marsalek, R.: Comparison of 802.11af and 802.22 standards-physical layer and cognitive functionality. Elektrorevue **3**(2), 12–18 (2012)
34. Leonardi, L., Patti, G., Lo Bello, L.: Multi-hop real-time communications over bluetooth low energy industrial wireless mesh networks. IEEE Access **6**, 26505–26519 (2018)
35. Li, Z., Uusitalo, M.A., Shariatmadari, H., Singh, B.: 5G URLLC: design challenges and system concepts. In: 2018 15th International Symposium on Wireless Communication Systems (ISWCS), Lisbon (2018), pp. 1–6

36. Lippuner, S., Weber, B., Salomon, M., Korb, M., Huang, Q.: EC-GSM-IoT network synchronization with support for large frequency offsets. In: IEEE Wireless Communications and Networking Conference (WCNC), Barcelona (2018)
37. Liu, Y., Kashef, M., Lee, K.B., Benmohamed, L., Candell, R.: Wireless network design for emerging IIoT applications: reference framework and use cases. Proc. IEEE **107**(6), 1166–1192 (2019)
38. Liva, G., Steiner, F.: pretty-good-codes.org: Online library of good channel codes. http://pretty-good-codes.org/
39. Liva, G., Gaudio, L., Ninacs, T.: Code design for short blocks: a survey. In: Proceedings of the EuCNC, Athens (2016)
40. Ma, L.: 5G Technologies, Standards and Commercialization. InterDgitial, Wilmington (2018)
41. MacKay, D.J.C., Neal, R.M.: Near Shannon limit performance of low density parity check codes. Electron. Lett. **32**(18), 1645–1646 (1996)
42. Mulligan, G., Bormann, C.: IPv6 over low power WPAN WG: IETF 73 (2008)
43. Page, J., Dricot, J.: Software-defined networking for low-latency 5G core network. In: International Conference on Military Communications and Information Systems (ICMCIS), Brussels (2016), pp. 1–7
44. Parvez, I., Rahmati, A., Guvenc, I., Sarwat A.I., Dai, H.: A survey on low latency towards 5G: RAN, core network and caching solutions. IEEE Commun. Surveys Tutorials **20**(4), 3098–3130 (2018)
45. Patti, G., Leonardi, L., Lo Bello, L.: A bluetooth low energy real-time protocol for industrial wireless mesh networks. In: IECON 2016 - 42nd Annual Conference of the IEEE Industrial Electronics Society, Florence (2016), pp. 4627–4632
46. Petersen, S., Carlsen, S.: WirelessHART Versus ISA100.11a: the format war hits the factory floor. IEEE Ind. Electron. Mag. **5**(4), 23–34 (2011)
47. Polyanskiy, Y., Poor, H.V., Verdu, S.: Channel coding rate in the finite blocklength regime. IEEE Trans. Inf. Theory **56**(5), 2307–2359 (2010)
48. Powell, M.: Bluetooth market update. Bluetooth SIG, Inc. (2018). www.bluetooth.com/wp-content/uploads/2019/03/Bluetooth$_$Market$_$Update$_$2018.pdf
49. Rappaport, T.S.: Wireless Communications: Principles and Practice, 1st edn. IEEE Press, Piscataway (2016)
50. Raza, M., Aslam, N., Le-Minh, H., Hussain, S., Cao, Y., Khan, N.M.: A critical analysis of research potential, challenges, and future directives in industrial wireless sensor networks. IEEE Commun. Surveys Tutorials **20**(1), 39–95 (2018)
51. Reed, S.R., Chen, X.: Error-Control Coding for Data Networks. Springer, Berlin (1999)
52. Ristiano, A.: ISA 100 Wireless: Architecture for Industrial Internet of Things. ETSI IEC 62734 (2014)
53. Sasaki, K., Makido, S., Nakao, A.: Vehicle control system for cooperative driving coordinated multi-layered edge servers. In: IEEE 7th International Conference on Cloud Networking (CloudNet), Tokyo (2018)
54. Schiessl, S., Al-Zubaidy, H., Skoglund M., Gross, J.: Delay performance of wireless communications with imperfect CSI and finite-length coding. IEEE Trans. Commun. **66**(12), 6527–6541 (2018)
55. Shannon, C.E.: A mathematical theory of communication. Bell Syst. Tech. J. **27**(3), 379-423 (1948)
56. Shirvanimoghaddam, M., et al.: Short block-length codes for ultra-reliable low latency communications. IEEE Commun. Mag. **57**(2), 130–137 (2019)
57. Siep, T.M., Gifford, I.C., Braley, R.C., Heile, R.F.: Paving the way for personal area network standards: an overview of the IEEE P802.15 working group for wireless personal area networks. IEEE Personal Commun. **7**(1), 37–43 (2000)
58. Sisinni, E., Saifullah, A., Han, S., Jennehag, U., Gidlund, M.: Industrial internet of things: challenges, opportunities, and directions. IEEE Trans. Ind. Inf. **14**(11), 4724–4734 (2018)
59. Sun, S., Fei, Z., Cao, C., Wang, X., Jia, D.: Low complexity polar decoder for 5G Embb control channel. IEEE Access **7**, 50710–50717 (2019)

60. Sutton, G.J., et al.: Enabling technologies for ultra-reliable and low latency communications: from PHY and MAC layer perspectives. IEEE Commun. Surveys Tutorials **21**(3), 2488–2524 (2019)
61. Van Wonterghem, J., Alloum, A., Boutros, J.J., Moeneclaey, M.: Performance comparison of short-length error-correcting codes. In: 2016 Symposium on Communications and Vehicular Technologies (SCVT), Mons (2016), pp. 1–6
62. Viterbi, A.: Error bounds for convolutional codes and an asymptotically optimum decoding algorithm. IEEE Trans. Inf. Theory **13**(2), 260–269 (1967)
63. Voigtlander, F., Ramadan, A., Eichinger, J., Lenz, C., Pensky, D., Knoll, A.: 5G for robotics: ultra-low latency control of distributed robotic systems. In: International Symposium on Computer Science and Intelligent Controls (ISCSIC), Budapest (2017)
64. Yang, Y., et al.: Narrowband wireless access for low-power massive internet of things: a bandwidth perspective. IEEE Wireless Commun. **24**(3), 138–145 (2017)
65. Zhang, L., Liang, Y., Xiao, M.: Spectrum sharing for internet of things: a survey. IEEE Wireless Commun. **26**(3), 132–139 (2019)
66. Zheng, K., Hu, F., Wang, W., Xiang, W., Dohler, M.: Radio resource allocation in LTE-advanced cellular networks with M2M communications. IEEE Commun. Mag. **50**(7), 184–192 (2012)

Part II
Automation Trends and Applications

Chapter 3
IoT-Driven Advances in Commercial and Industrial Building Lighting

Daniel Minoli and Benedict Occhiogrosso

3.1 Introduction

The Internet of Things (IoT) is penetrating the diurnal operation of many business and industrial sectors. Some refer to the IoT as being comprised of "cyber-physical systems", or as being the process of "digitalization of physical surfaces". Early and tangible applications of IoT will be in the smart building and in the smart city contexts.

Smart building IoT systems entail the ability to monitor, analyze, and control a broad range of systems and system parameters related to ambient temperature (especially in conjunction with legacy Building Management Systems [BMSs]), lighting, occupancy, security, behavior (e.g., bans on vaping or smoking on premises), and comfort control (e.g., indoor air quality, space and bathroom cleanliness). Ambient temperature and lighting have a direct impact on energy consumption, and energy management is important from both a financial perspective and a regulatory perspective. The need to manage energy consumption is well appreciated by facility management stakeholders and government entities [1]. Some localities have a goal of achieving a city carbon neutrality by 2050—e.g., recently New York City adopted legislation to reduce carbon emissions in buildings larger than 25,000 square feet by 20% between 2020 and 2030 and 80% by 2050, and smart lighting can make a significant contribution towards that goal [2]. Office space occupancy relates to the efficiency of space utilization, and thus, also has a financial implication, since urban office space can cost up to $75–100/square foot/year. Physical security is obviously a requirement that needs to be satisfied; behavior and comfort are also requirements but may be at a lower level of business priority. Industrial (smart) factories also fit the rubric of the "smart buildings" paradigm.

D. Minoli (✉) · B. Occhiogrosso
DVI Communications, New York, NY, USA
e-mail: daniel.minoli@dvicomm.com; ben@dvicomm.com

I. Butun (ed.), *Industrial IoT*, https://doi.org/10.1007/978-3-030-42500-5_3

Smart City (IoT) systems aim at improving the efficiency of city asset management (e.g., mobility, vehicular automation, traffic/traffic lights control—such as with Intelligent Transportation Systems—parking, multi-modal transportation, and "smart street" services); smart power and utility grids; management of street lighting and other municipal functions such as waste collection; water management; monitoring of the environment (e.g., using sensors on institutional vehicles to collect data for environmental phenomena); surveillance/intelligence (e.g., facial recognition); smart business services (e.g., location-based services); Quality of Life (QoL) of the residents, crowd-sensing (where the inhabitants of the city or region use smartphones, wearable devices, and car-attached sensors to collect and upload a plethora of optical, signal, and ambiance data); smart government; and, enhanced connectivity such as provided by 5G cellular services (and the ensuing micro-tower "densification").

While each of the application areas identified so far is of great interest, this chapter assesses technical, economic, and market aspects of *evolving lighting technologies* in the context of smart buildings, smart factories, and smart cities. Light Emitting Diodes (LEDs) are semiconductor diodes operating on low-voltage DC currents that can be controlled digitally. The combination of LEDs as light sources with microcontrollers achieves a good level of device intelligence, making these sources *bona fide* IoT entities, and enables the luminaires (lighting fixtures) to be controlled using various types of protocols and network connections. LED technology has greatly reduced the amount of power required for lighting applications at a given level of luminance. As early as 2013, the assertion was made that "the lighting industry has been going through a fundamental digital revolution: with light sources going for LED, drivers going digital, and control going networked" [3]. In this review and application chapter the following topics are discussed:

- Commercial Real Estate Smart Lighting (building applications);
- Commercial Real Estate Power Over Ethernet (PoE) LED-based Lighting (building applications);
- LED-based (Smart) Street lighting (city application).

It is acknowledged that LED-based street lighting can support additional services (e.g., Wi-Fi amenity hotspots, video surveillance) based on unlicensed frequency bands (for example at the 2.4 GHz band), or based on licensed services, specifically 5G microcells to support traditional connectivity as well as new IoT services (such as NB-IoT or LTE-M). The emphasis of this chapter, however, is on commercial real estate and industrial lighting.

For industrial applications Smart Lighting will improve and optimize the work environment, including energy efficiency and biological benefits for second and third shift workers by creating appropriate lighting conditions as well as the use of personnel vacancy sensors to "work lights out" in highly automated factories in the future.

Section 3.2 provides motivations for a conversion to next-generation lighting. Section 3.3 reviews fundamental concepts of lighting, biological basics and IoT concepts. Section 3.4 covers applicable technologies and standards including LED technology, lighting controls, lighting control connectivity and networking, ZigBee, Bluetooth, Wi-Fi, Li-Fi, LoRa and Sigfox, NB-IoT and LTE-M, wired systems, PoE technology, PoE lighting, and lighting standards. IoT support of lighting is discussed also in conjunction with Building Management Systems. Outdoors/Street lighting concepts are also covered. Section 3.5 briefly assesses some market considerations. Design considerations, examples of costs, implementation, and operational considerations are highlighted in the remaining sections.

3.2 Motivations

Unlike the typical mantra where a "technology solution is in search of a problem" energy efficiency is a "problem in search of a technical solution". Principally, energy efficiency saves corporate and institutional money to the bottom line, but, as noted, there are also regulatory requirements that aim at controlling carbon emissions to address long-term climate considerations.

The Commercial Buildings Energy Consumption Survey (CBECS) is a U.S. national survey collecting information on the inventory of commercial buildings, also including energy usage data (consumption and expenditures). Their data shows that there were 5,600,000 commercial buildings in the U.S. in 2012, spanning 87.4 billion square feet of floor space [4]. At this juncture, 2012 is the most recent year for which the survey data is available; data collection for the 2018 CBECS started in early 2019 and preliminary results will be available to the public in the Spring of 2020 [5]. Worldwide *buildings* (commercial and residential) account for over 40% of total energy consumption. Per reference [4], office buildings consume an average of 15.9 kilowatt-hours (KWh) of electricity per square foot annually, which accrues to an annual expenditure of US $1.70 per square foot (employing a typical figure of $0.106/KWh). For the average office building in the U.S. (which has an area of 15,000 square feet), electricity consumption equates to $25,500 annually. Assuming that the rent cost (say at $60/year/square foot) equates to $900,000 per year, the energy costs, which are often, but not always, incremental to the rent, are about 3% compared to the rent disbursements. Information from reference [4] also shows the following details related to *electricity* usage (in recent times):

1. When looking at all buildings, the electrical energy consumption is as follows:
   - Space heating at 2.0%; *cooling at 14.9%; ventilation at 15.8%*; water heating at 0.5%; *lighting at 17.1%; cooking at 2.2%*; refrigeration at 15.8%; office equipment at 4.1%; computers at 9.5%; and other at 18.1%.
   - Therefore, 32.7 % of the electrical energy is consumed by space heating, cooling, and ventilation; the next big item is lighting at 17.1%; office equipment and computers account for 13.6% (for a total of 63.4%).
2. When looking just at *office* buildings (they consume 20.1%of the total U.S. electrical energy use), the energy consumption is as follows:

- Space heating at 2.2%; *cooling at 13.4%; ventilation at 24.7%;* water heating at 0.2%; *lighting at 17.1%;* cooking 0.2%; refrigeration at 3.2%; office equipment at 4.3%; computers at 19.3%; and other at 15.3%.
- Therefore, 40.3% of the electrical energy is consumed by space heating, cooling, and ventilation; 17.1% for lighting; office equipment and computers account for 23.6% (for a total of 81.0%).

Other estimates place the total energy consumption for commercial and institutional building lighting and for street lighting at 11% of the total U.S. consumption [6]. (Comparable energy allocation data applies to many other industrialized nations). When also considering residential usage, lighting accounts for 15% of the global electricity use and it accounts for 5% of the global greenhouse gas emissions [7]. In the U.S. about 63% of the electricity is generated by fossil fuels (30% from coal) [8].

The mission of a lighting system is to provide illumination where it is needed, when it is needed, and in the intensity that is needed for the environment or task at hand. Appropriate control and specifically pertinent control algorithms are required to achieve the illumination required while maximizing energy savings [9]. In the sections that follow, the baseline average office space parameter is utilized; for different size offices, the data/results can be scaled linearly. It should also be noted that the usable/lit space in an office environment is typically 85% of the nominal "lease" footage. As further motivation a worldwide survey of several hundred commercial real estate developers, owners, and tenants recently undertaken by Harvard Business Review Analytic Services indicated that 66% of these organizations identified energy management as the fundamental business driver for smart buildings [1]. Additionally, the survey showed that 72% of executives surveyed stated that their principal business goals for smart buildings are reducing facilities and operations costs, thus improving bottom-line profitability.

Smart Lighting (also known as "network-connected lighting", "connected lighting", or "digital lighting") facilitates significant savings and operational improvement. To start with, LED lighting technology can reduce energy consumption in the range of 50-70% compared to traditional lighting. Therefore, there is a clear motivation to undertake a conversion. At a macro level, a conversion to the new LED/IoT/(PoE) lighting systems could in theory save the average office of (net) 15,000 square feet from $15,000 × 1.7 × 0.5 = $12,750 to $15,000 × 1.7 × 0.7 = $17,850 a year in electricity costs. The payback—considering the project conversion costs—ranges from 3 to 5 to 7 years, depending on various circumstances. Beyond this simple LED conversion, connected lighting in building environments (for LED or traditional luminaires) aims at improving lighting controls, which by itself can result in saving 30–40% of the energy use through occupancy sensing, daylight harvesting, dimming, and other optimization mechanisms [10]. However, there are technology choices, product openness, cost assessment, implementation (project management/project delivery), control, and reliability, integration, and cybersecurity considerations to be assessed in effecting such conversion.

There is a need for intelligent lighting management for commercial buildings that facilitates improved energy savings, easy maintenance, and simple implementation—whether as an initial deployment or a retrofit—and optimal control. Evolving light management systems support these benefits. First-generation connected lighting systems (CLSs) faced some of the same challenges of traditional lighting controls: they were complicated and difficult to configure and commission. Second-generation systems aim at addressing these limitations. For building applications, there are fundamentally two scenarios: new construction and retrofit; different cost/implementation considerations apply to each case. Residential consumers are also adopting smart light fixtures, given that these allow one to easily change the ambience of a home by dynamically altering the color temperature. There is also a trend for support of voice controls from virtual assistants such as Amazon Alexa, Siri, Google Assistant or Apple HomeKit.

3.3 Fundamental Concepts

3.3.1 Lighting Concepts

Humans perceive a segment of the electromagnetic spectrum as visible light. The wavelength that spans visible light runs from approximately 360 nm to 770 nm, with the low end corresponding to violet light and the high end corresponding to a (deep) red light. See Fig. 3.1.

The Spectral Power Distribution (SPD) is a plot of the radiant power emanating from a source (specifically, a light source) at each wavelength in the visible region of the spectrum [11]. It is measured as radiant power (μW/10 nm of spectrum/lumen) versus spectrum. It is a measurement that expresses the power per unit area per unit wavelength of a light source (radiant exitance or radiant emittance). Light energy can be precisely characterized by defining the power level of the light at each wavelength in the spectrum. Such a plot either shows actual power per nm, or it depicts relative intensity in arbitrary units. In a laboratory setting the SPD can be established by a utilizing spectrophotometer.

Figure 3.1 also depicts a SPD graph of a light profile known as D65. SDP concepts are discussed further below, but a short description of the Commission-Internationale-de-l'Eclairage (CIE) Standard Illuminant D65 follows herewith: D65 is a known standard illuminant that endeavors to portray open-air conditions of direct sunlight with the light diffused by a clear sky, specifically approximating the average noon light in Western Europe; it is represented as a table of averaged spectrophotometric data, specifically by a given SPD. Thus, D65 profile aims at representing light in an average day and has a Correlated Color Temperature (CCT) around 6500 °K.

Figure 3.2 provides examples of SPD; it depicts a continuous power spectrum (for example for a 2800 °K incandescent house lamp) along with the spectrum of a

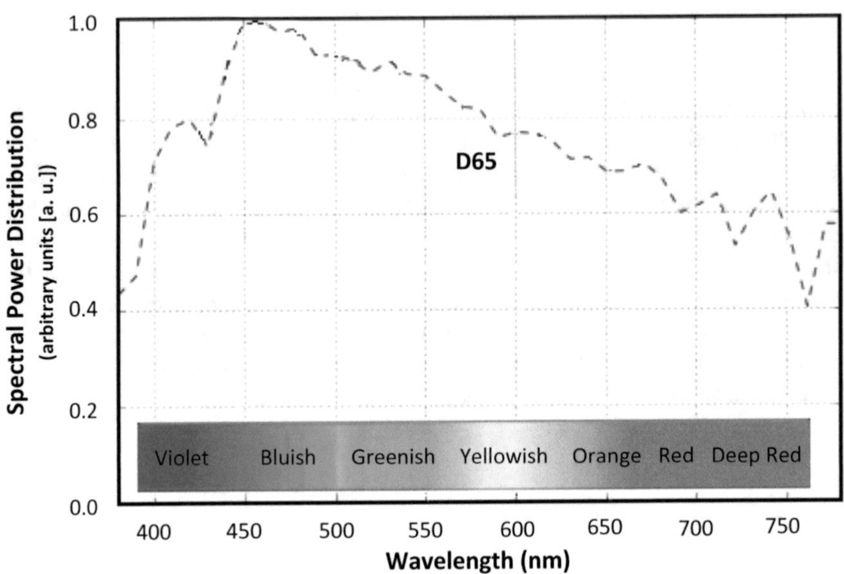

Fig. 3.1 Visible light spectrum (representative profile)

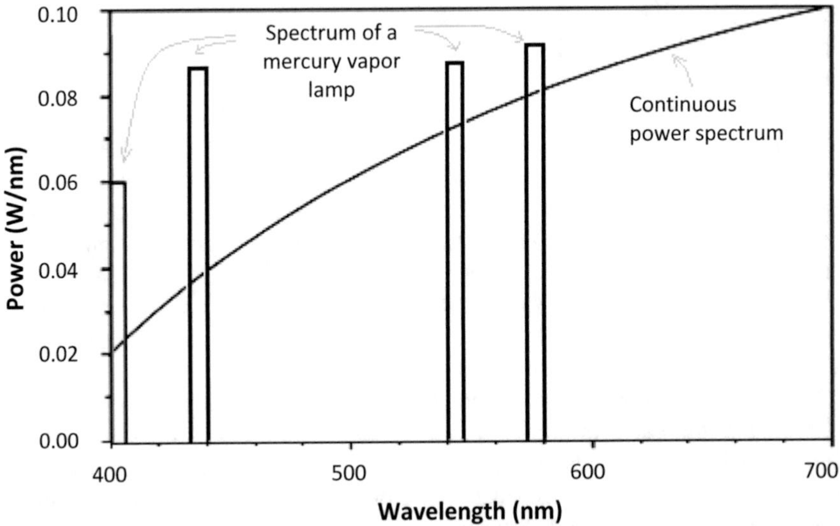

Fig. 3.2 Spectral power distribution (generic)

mercury vapor lamp—an incandescent lamp produces a continuous spectrum, while a mercury vapor lamp emits light at a handful of primary wavelengths. The spectral profile of a light source affects its ability to render colors in a "natural" manner. Luminaire manufacturers often publish SPD graphs of their various light sources.

Sometimes the concept of a *Relative SPD* is utilized. This is the ratio between the spectral concentration at a given wavelength and the concentration at a reference wavelength.

The variable light color of luminaires can be defined by the hue, saturation and brightness. Light sources vary in their ability to deliver the color of objects in a "natural manner." An important concept, in this context, is the Color Rendering Index (CRI). CRI is a quantification of the extent of color alteration that an object undergoes when illuminated by a particular light source as compared to the color of the object when illuminated by a reference light source having a CRI of 100 and equivalent color temperature. Another important concept is the CCT which is a description of the color appearance of the light generated by a particular light source, relating its color to the color of a reference light source when heated to a specified temperature measured in degrees on the Kelvin (K) scale. The CCT rating of a light source (e.g., a lamp) is a proxy for its "coolness" or "warmth". For example, a light source with a CCT rating between 2000 °K and 3200 °K (e.g., a High Pressure Sodium source, a Tungsten Halogen source) produces light which ranges from orange to yellow white) and are perceived as "warm", while those with a CCT above 4000 °K (e.g., Linear Fluorescent) (in the blueish range) are perceived as "cool". Table 3.1 provides some key concepts related to lighting.

In the context of LEDs, combining signals with given wavelengths generated by LEDs with (algorithmically) calculated amplitudes can produce light that appears to be white. Thus, it is possible that the light from two lamps can have different wavelength combinations and yet appear to be of the same color (i.e., same nominal CCT); the effects of these two lamps on objects may be very different (as measured by the CRI) [11–13]. White tuning relates to a source's ability to shift its CCT; this is typically done by proprietary dimming algorithms implemented in the on-board lamp or in the lighting engine. Approaches include warm dimming (as the intensity decreases the CCT gets warmer), tunable white (independently controlling the CCT and the intensity), and spectral tuning (sophisticated wavelength processing). Note that in dimming one must distinguish between the measured light—the amount of light registered on a light meter—and the perceived light—the amount of light that the eye interprets; for example, after dimming, if the measured light is 20%, the perceived light would be 45%.

There are a number of standards that define color (chromaticity) ranges for white light LEDs with various CCTs, and other parameters, including but not limited to the following U.S. standards:

- American National Standards Institute (ANSI)/ National Electrical Manufacturers Association (NEMA) spec C78.377.2008 [14];
- ANSI/NEMA spec C82.77-2002 [15];
- Illuminating Engineering Society of North America (IESNA) spec TM-16-05 [16];
- IESNA spec LM-80-08 [17].

Table 3.1 Basic light-related concepts

Term	Description
Luminous flux (Φ_V)	The part of the power (W) that is sensed as light by the human eye, expressed in lumens. Total radiated power (Φ) in W while considering the sensitivity of the human eye. It is the energy per unit time being radiated from a light source over the human-visible wavelengths (from about 380–780 nm) given that the eye does not react uniformly to all optical wavelengths, the luminous flux is taken as a weighted average of the radiant flux
Lumen (lm)	The unit for the luminous flux of a light source, as defined by International System of Units (also known as SI). It represents the luminous flux emitted into a unit solid angle by an isotropic point-source which has a luminous intensity of 1 candela. Simply, luminance light emitted by or from a surface, and a lumen is quantity of visible light emitted from a source. Luminous flux (lm) = Radiant power (W) * 683 lm/W * luminous efficacy of the light For example, a 100-W incandescent lamp has a luminous flux of about 1700 lm, or 17 lm/W
Luminous efficacy (V_λ)	A quantitative assessment of how well a light source produces visible light; it corresponds to the ratio of luminous flux to power and is measured in lumens per W. A weighting factor that enables the conversion of Radiant Flux (Φ) to Luminous Flux (Φ_V). In the photopic region, the peak at 555 nm is given a conversion value of 683 lm/W $\Phi_V = \Phi * V_\lambda * (683 \text{ lm/W})$ Example 1: a 10 mW laser pointer using a specific wavelength of 680 nm produces 0.010 W * 0.017 * 683 lm/W = .116 lm Example 2: a 10 mW laser pointer using a specific wavelength of 630 nm produces 0.010 W * 0.265 * 683 lm/W = 1.81 lm Example 3: The luminous flux from a source radiating energy over a spectrum is more complex; one must utilize the SPD for the source in question; one then calculates the luminous flux at defined intervals for continuous spectra, and summing up the flux at each interval will generate a total flux produced by a source in the visible spectrum. Generally, the manufacturer gives the luminous flux for a lamp
Reflectance	The quantity of light a surface reflects compared to the amount that hits the surface (often expressed as a percent)
Correlated color temperature (CCT)	A description of the color appearance of the light generated by a light source, relating its color to the color of reference light source when heated to a specified temperature. This rating is a proxy for a source's "warmth" or "coolness", as follows: Less than 3500 °K: Warm; 3500–4000 °K: Neutral; 4100–5000 °K: Cool; above 5000 °K: Daylight
Typical units of light measurements	Foot-candle (fc): a unit used to quantify light intensity or illuminance. 1 foot-candle (fc) = 10.76 lux Foot-lambert (fl): a unit of brightness or luminance. It equates to reflectivity (as a percentage) * illuminance (in fc) Lux: Is a light measurement unit, equal to one lumen per square meter. 1 lux = 0.09 fc
Spatial daylight autonomy (sDA)	A measure of illuminance sufficiency during the day for an area under consideration, documenting the percentage of the floor area that exceeds a specified illuminance (e.g., 400 lux) for a specified percentage of the analysis period

3.3.2 Biological Basics

The color temperature of emitted light is believed to impact human productivity and can be tuned in accordance with schedules tied to circadian rhythm. Humans experience three visual states (see Fig. 3.3):

- *Photopic vision*: experienced in well-lit space during the day. Color perception and visual acuity are optimal in this state. Cones in the retina are most active in this state. During the day, people are most sensitive to the green-yellow portion of the spectrum (555 nm). The luminance is generally 3 foot-candles per m^2 or higher (30 lux). 555 nm represents the peak efficiency of the photopic (daylight) vision curve.
- *Mesopic vision*: experienced at dusk, when details start to, but not completely, wane. Both rods and cones participate in perception in this state. The luminance generally ranges from 0.01 to 3 foot-candles per m^2.

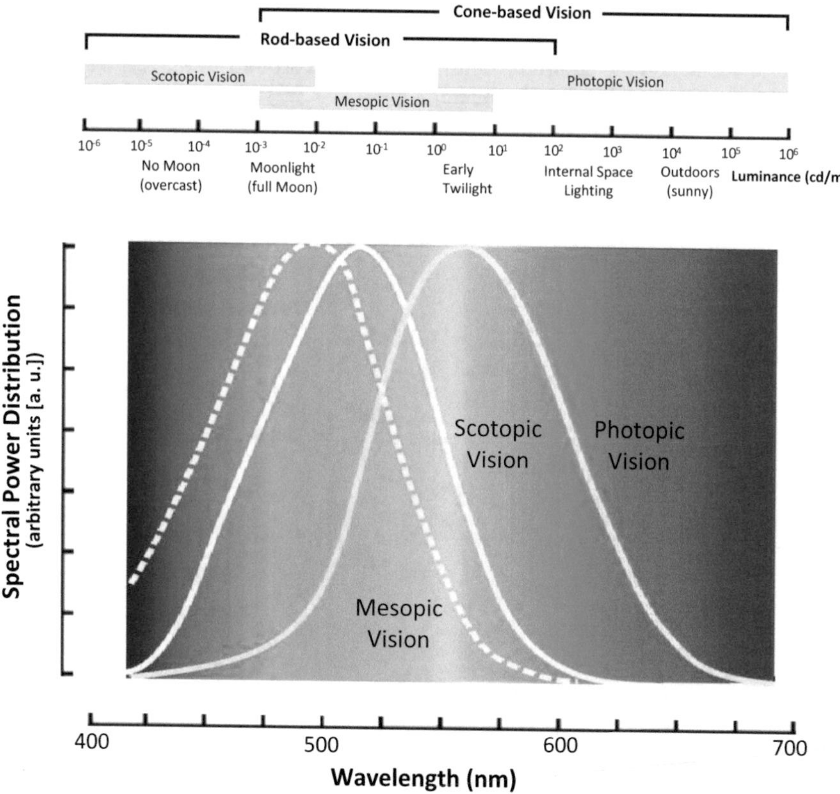

Fig. 3.3 Human vision states

- *Scotopic vision*: experienced when it gets dark. Rods in the retina are most active in this state. People lose color discernment but retains some sensitivity in the green-blue region (498 nm). The luminance is generally below 0.01 foot-candles per m^2, thus visual acuity will wane.

The daylight sensitivity (photopic vision sensitivity) of the eye peaks at 555 nm and falls off to practically zero at 380 and 750 nm; the eye's nighttime sensitivity (scotopic vision sensitivity) shifts toward the blue, peaking at 507 nm falls off to practically zero at 340 and 670 nm.

The melanopic segment of the visible spectrum (supporting the mesopic vision) defines the frequency range where the melanopsin-containing cells in the retina react to the stimulus of light energy (see Fig. 3.4). Melanopic lighting is lighting that is optimally tuned to both non-visual and visual responses to light by humans. In fact, for humans there is more than just visual perception, there is, additionally, a biological impact of light: the hormone melatonin (from which the "melanopic" originates) is a modulator to people's pattern of sleep and wakefulness (this being the circadian rhythm). The production of the melatonin hormone is suppressed during the day by a naturally blue-rich light environment; people tend to fall asleep at night when there are little blue hues in the light. As seen in Fig. 3.4, the spectral distribution at which melatonin is controlled peaks at around 464 nm, a wavelength that corresponds to a strong blue. Manufacturers of LEDs seek to adjust the light output of their sources to produce artificial light sources that helps people to retain their attention to detail at times when they may otherwise be slumping (melanopic

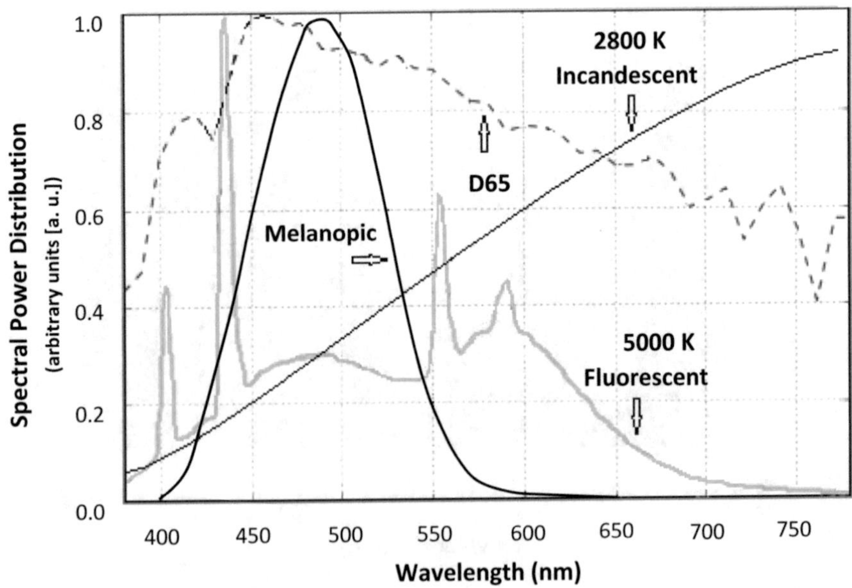

Fig. 3.4 Spectral power distribution (examples)

lighting may also assist shift workers in offices or industrial complexes to remain productive). LED technology enables fine adjustments (with appropriate algorithms in the control microprocessors [CMPs] and/or drivers) to the (white) light, to achieve warmer tones also combining the melanopic peak within the produced spectrum.

Algorithms are utilized to manage energy consumption and shape the LEDs' operation to achieve the desired light spectral profile. Note that Fig. 3.4 also shows an example of an incandescent light source; typically, such a source generates more spectral power at the longer wavelengths (>650 nm), thus, it renders red colors of an illuminated object quite effectively. In this figure the fluorescent lamp has more spectral power in the short wavelength (<450 nm); thus, the blue colors of an illuminated object appear very vivid.

The circadian system regulates physiological rhythms in the human body, including organs and tissues, affecting the levels of hormone along with the sleep-wake cycle. Circadian rhythms are synchronized by a number of cues, including light energy stimulating intrinsically- photosensitive-retinal-ganglion-cells (ipRGCs). ipRGCs are the eyes' non-image-forming photoreceptors: ipRGC are a non-visual pathway that is involved in the circadian/melanopic cycle) [18]. Light energy at low wavelength (high frequency) and high intensity boosts alertness, while the dearth of this energy signals the body to reduce energy output and get set for rest. The biological effects of light and lighting can be quantified in Equivalent Melanopic Lux (EML), focusing on the ipRGCs – the traditional focus is on the eye cones, whose response is measured with traditional lux metrics. A typical design goal for melanopic light intensity for work areas is: for at least the hours between 9:00 in the morning and 1:00 in the afternoon, for every day of the year, for a least 75% of workstations, at least an EML of 200 is available, measured on the vertical plane facing forward, 4 ft. above the floor (to simulate the view of the occupant). The lights under assessment may be dimmed in the presence of daylight energy but should be able to independently achieve these levels. Other guidelines are available for where office or factory workers spend most of their time in spaces with light levels limited by work type (e.g., hospital rooms, assembly lines) or other conditions (e.g., primary and secondary schools) [18].

Effective indoor lighting is achieved by providing uniformly-distributed light with stable light output and appropriate color temperature. The human eye is sensitive to fluctuating light intensity (especially with a flicker rate below 75 Hz), at higher optical signal wavelengths (lower frequencies, e.g., in the orange/red zone); fluctuating light intensity can give rise to headaches and eye irritation. LED lamps are subject to AC line-related fluctuations and random (stochastic) light intensity fluctuation. The former has a frequency at double the line frequency 100/120 Hz for 50/60 Hz AC line; the latter is typically caused by component incompatibility. Light flicker must be properly controlled. Lighting control not only enables the lighting to be adjusted to suit the visual requirements but also allows it to shape and interpret the architecture of the locale in question (e.g., office, home, retail store, theater, museum) [19].

Additionally, beyond proper lighting, office wellness includes air quality, water quality, thermal comfort, sound and noise, space adequacy for the work function at hand, and optimal movement facilitation.

3.3.3 IoT Concepts

Typical functionality of IoT includes embedded intelligence, sensing (capturing data), data analysis at the edge, device and tag connection (data transfer), and analytics (big data analysis, AI). Numerous definitions and descriptions of the IoT exist (e.g., certainly not limited to [20–28]). One fundamental/descriptive quote from the author's previous work [29] is as follows: "*The basic concept of the IoT is to enable objects of all kinds to have sensing, actuating, and communication capabilities, so that locally-intrinsic or extrinsic data can be collected, processed, transmitted, concentrated, and analyzed for either cyber-physical goals at the collection point (or perhaps along the way), or for process/environment/systems analytics (of a predictive or historical nature) at a processing center, often 'in the cloud'. Applications range from infrastructure and critical-infrastructure support (for example smart grid, Smart City, smart building, and transportation), to end-user applications such as e-health, crowd sensing, and further along, to a multitude of other applications where only the imagination is the limit*". The value of IoT derives from the information one can capture and the actionable analytics that comes from such aggregated data. Some well-known protocols of an IoT protocol stack are shown in Fig. 3.5 (other vendor- or organization-specific protocols also exist, e.g., but not limited to DASH7, Z-Wave, INSTEON, EnOcean).

IoT-based applications include, among others, smart city, smart campus, smart building, healthcare, transportation, industrial process control—many of these also entail occupancy and positioning. Sensors include devices that address the following among others: ambient light, position, presence, proximity, motion, acceleration, tilt, force, torque, pressure, velocity, displacement, sound, vibration, temperature, humidity, gases, ambient conditions, magnetic fields, optical surveillance video and frame grabs.

In the context of IoT, lights in commercial real estate are interconnected with power lines, and in a number of instances, with dedicated data lines to transfer information to a control system; other control approaches are wireless. Lights have and/or can accommodate onboard sensors and communication transceivers (wired or wireless), thus they fit the definition of IoT and exist and are part of an IoT ecosystem (such ecosystem may include BMSs, surveillance and access systems, and flows related to other mechanical systems). The IoT enables the full realization of the Smart Building concept, with lighting being one (key) component. Additional characterizations follow.

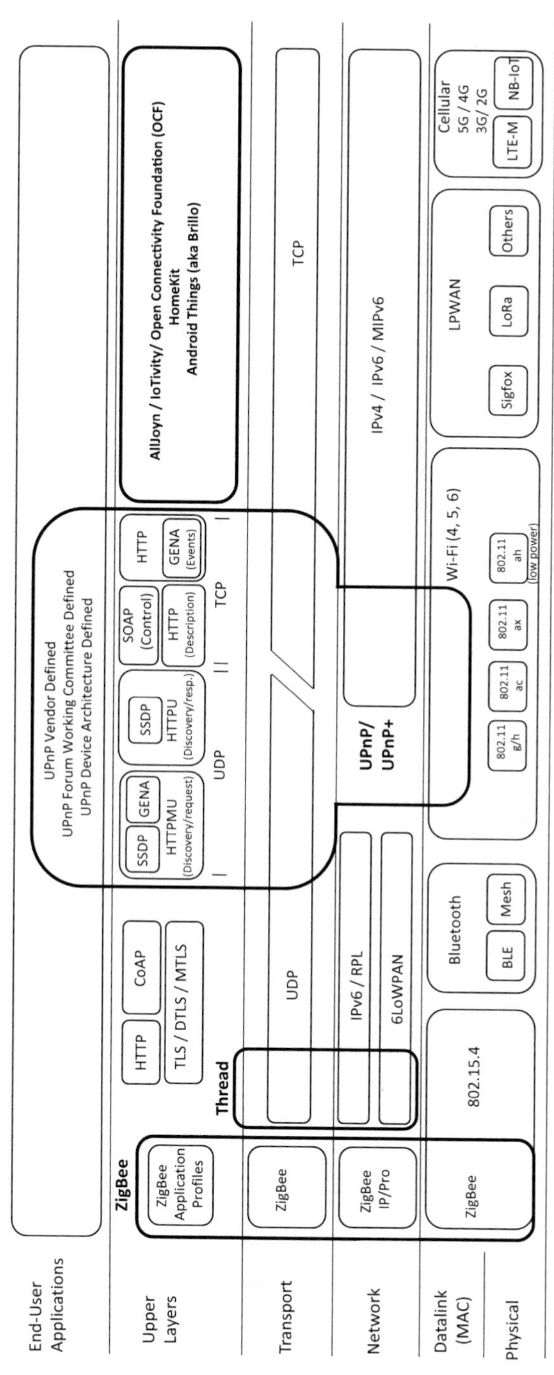

Fig. 3.5 Common IoT protocol stacks

3.4 Technologies and Standards

The fundamental function of a lighting system is to deliver a specified amount of light to a space or an environment consistent with environment-appropriate design criteria while, simultaneously minimizing energy consumption. Lighting control approaches based on occupancy and daylight adaptation, now common in building energy codes, can materially lower lighting energy consumption compared to fixed systems; however, more sophisticated controls, and also access to analytics, are desired by facilities managers in commercial and industrial settings [6]. As noted in the introduction, advancements in lighting can be grouped into two major categories:

- *Smart Lighting* ("network-connected lighting" or "connected lighting" or "digital lighting" or "Digital ceiling"), which, almost invariably, utilizes LED luminaires and IoT-based sensors to support a number of control functions (the sensors can be wired or wireless). These systems are collectively known as Connected Lighting Systems (CLSs). Some additional non-lighting sensors can also be deployed as part of a smart lighting deployment, if desired, e.g., Wi-Fi Access Point or licensed services, specifically 5G microcells or Distributed Antenna Systems (DAS); these additional functions are currently more typical in smart city street lighting applications than in smart building applications.
- *PoE LED-based Lighting*, in commercial real estate buildings both the controls and the power to certain lighting fixtures are based on PoE, occasionally resulting in lower installed cost.

Connected (networked, smart) lighting systems, CLSs, are fostering a transformation of the building's informatics landscape by providing a data platform supporting cohesive building resource management. Smart Lighting naturally addresses human factors and economics, but the fundamental concepts (supported by IoT principles) are control, sensing, and connectivity. Sensing is the first step needed for control. Sensing can include (but is not limited to) color temperature, luminance (lux level), occupancy, daylight and ambient light, but can also include non-illumination-related sensors such as air quality, and temperature and humidity. Human factors include human behavioral preferences, color perception, and circadian cycles discussed earlier. Some of the preferences are achieved by making use of a large set of different luminaires and fixtures, each adapted to the space at hand (e.g., office, conference room, corridor, open spaces). Economics deal with energy usage in the LED context (also in conjunction with the control for daylight harvesting, occupancy, seasonal variations), and also with the variety of available luminaires and fixtures to cost-effectively light the (heterogenous) office space environment.

Connectivity in support of intelligent control can be wired or wireless. Wired systems include, among others, Power-line Communication (PLC), RS-485, Local Operating Network (LON), Konnex (KNX), Digital Multiplex (DMX), and Digital Addressable Lighting Interface (DALI). Common wireless links include but are not limited to ZigBee Light Link (ZLL), Bluetooth Mesh Network, Wi-Fi, Li-Fi, and also, possibly LoRa and Sigfox, and NB-IoT and LTE-M.

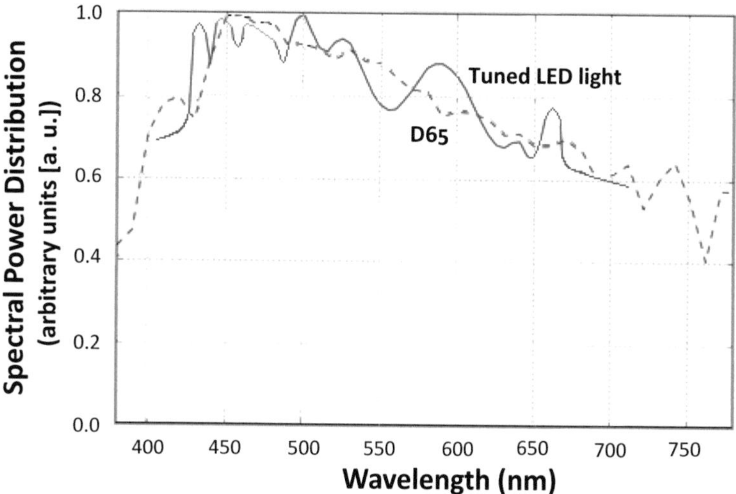

Fig. 3.6 Example of LED tuning to emulate D65 light

For indoors applications, key control parameters (for both Smart Lighting and PoE-based lighting) include luminance levels, color temperature, presence, fixture status; other capabilities (in conjunction with add-on components) might include air quality monitoring, window shades controls, dispensers, and so on. In addition, control deals with data collection and analysis, algorithms (tunable light sources necessitate non-trivial algorithms to generate target visual profiles—e.g., tuning the weights of, say, 10 different channels of LED luminaire—see Fig. 3.6 for an example), looped systems, and energy audits. For outdoor applications, key control parameters include dual intensity luminance levels with traffic-presence sensors, and intensity luminance levels with weather/moonlight/seasonal capabilities; additional features related to the control of Wi-Fi/5G services, if provided, and surveillance modalities if provided.

3.4.1 LED Technology

A LED is a semiconductor device that generates light when electrical current traverses it in one direction. Multi-channel LED systems can create light in the visible range (380–780 nm). In this context the term "LED" refers to both an LED (retrofit) lamp or to an LED fixture (luminaire). LEDs and organic light-emitting diodes (OLED) lighting are examples of Solid-State Lighting (SSL) technologies. SSL lighting systems are comprised of light sources, solid state drivers, digital sensors, CMPs, and communication interfaces (wired or wireless) (see Figs. 3.7 and 3.8). LED emitters are inexpensive and are easy to fabricate, being comprised of

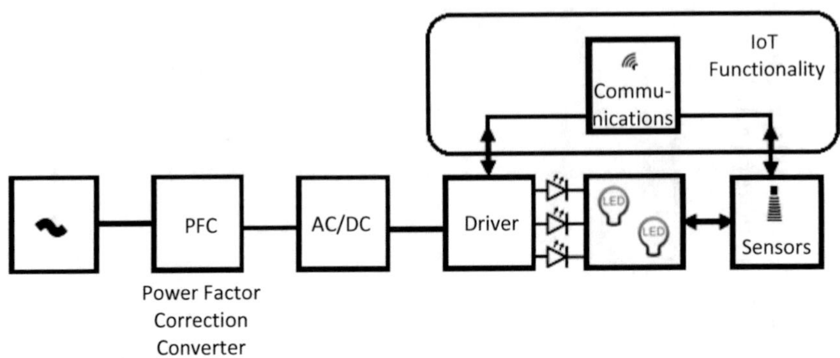

Fig. 3.7 LED IoT-enhanced LED lighting system

non-polluting materials. Efficacies reaching up to 250 lm/W can be achieved; LED-based fixtures are more cost-effective overall as compared to traditional lighting systems, especially considering that extended lifespans with more than 100,000 h can be typically achieved. LED light sources consume less energy than traditional lighting from incandescent bulbs, fluorescent tubes, halogen lamps, or mercury vapor lamps.

LED drivers are devices that convert the line-voltage power (e.g., 120/220/277 V AC) to the low voltage used by the LEDs. Drivers typically also interpret control signals to dim or otherwise control the LEDs. LED drivers can achieve high conversion efficiencies ranging between 85% and 95% over a wide range of input voltages. LED drivers operate in either constant current or constant voltage; they are not interchangeable. Both LED lamps and LED fixtures require LED drivers. Constant voltage drivers furnish a fixed constant voltage to LED modules connected in parallel; this approach is typically used in areas where one may have a variable number of fixtures. Constant current drivers furnish a constant current to a specific LED module that is designed to operate at that current level; this approach is typically used for LED fixtures that utilize only one LED module per driver (this being comparable to a fluorescent lamp with its dedicated ballast) [30].

LEDs that are integrated into a variety of fixture models meet the illuminance needs of office space while at the same time increasing visual comfort and reducing energy consumption. A CCT between 2700 °K and 6500 °K and Color Rendering Index (CRI) up to100 can be achieved. LED luminaires typically include advanced spectral features for dynamic white light reproduction and color mixing (white light reproduction employs a combination of RGB LEDs or, alternatively, can employ a phosphorous coating in conjunction with LED encapsulation resin). LEDs are not intrinsically "white light" sources; the majority of the "white light" LED products on the market use the phosphorous conversion method, where a blue (or near-ultraviolet) LED diode is coated with phosphor (cerium-doped Yttrium aluminum garnet (YAG)) to enable the assembly to emit what to the human eye appears to be white light [31]. LED color and materials for each color are as follows: red

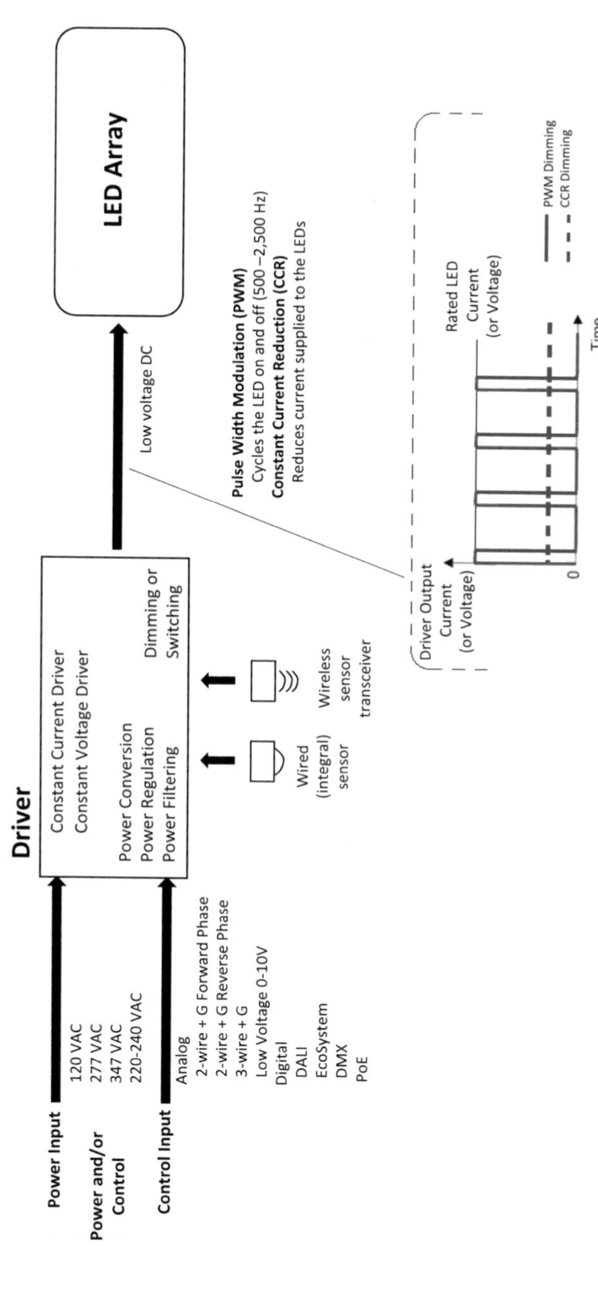

Fig. 3.8 LED driver functions

with Aluminum Gallium Arsenide (AlGaAs); green with Indium Gallium Nitride (InGaN); and blue with Zinc selenide (ZnSe); the relative intensity of each LED emissions will result in different "types" of white such as white warm, white neutral, or white cold. The phosphorous coated LEDs entail inorganic blue LEDs made of InGaN with phosphorous coating made of Yttrium Aluminum Garnet (YAG) synthetic crystalline materials.

LED emitters can be modulated at high speeds in the range of several MHz (also allowing applications for local optical communications such as Li-Fi and Visible Light Communications [VLC]). LEDs can be classified based on input current requirements as high-power LEDs (more than 150 mA), mid-power LEDs (30–150 mA) and low-power LEDs (5–20 mA). High-power LEDs are typically employed in illumination applications (also in automotive applications); mid-power LEDs are useful in Li-Fi and VLC communications, and low-power LED emitters are utilized for display applications. The efficacy of commercial LED luminaires typically ranges from 90 to 120 lm/W, however, for residential applications the efficiency of bulbs is in the range of 40-70 lm/W. See Fig. 3.9 (also showing PoE power inputs for comparison).

Technological advancements in LED systems in recent years are making this technology very attractive to the lighting world. Table 3.2 provides a simple comparison between Incandescent systems and LEDs. The expectation has been that LEDs will account for 69% of light bulbs sold and over 60% of the installed global base by the end of 2020 [32]. The U.S. Department of Energy (DoE) forecasts that by 2035, LED lamps and luminaires will comprise 86% of installed stock (compared with just 6% in 2015) and, compared to a no-LED scenario, the annual energy savings from LED lighting will represent a 75% reduction in consumption [33]. LED lightbulbs reduced 570 tons of carbon emissions in 2017 (equivalent to 162 coal-fired plants) [34].

Recent developments in LED technologies include OLEDs, microLEDs, and phosphorous coated LEDs. OLED are LEDs where the electroluminescent medium is a film of an organic compound that is able to emit light in response to a current; OLEDs are used as displays in devices such as TV screens, but there is

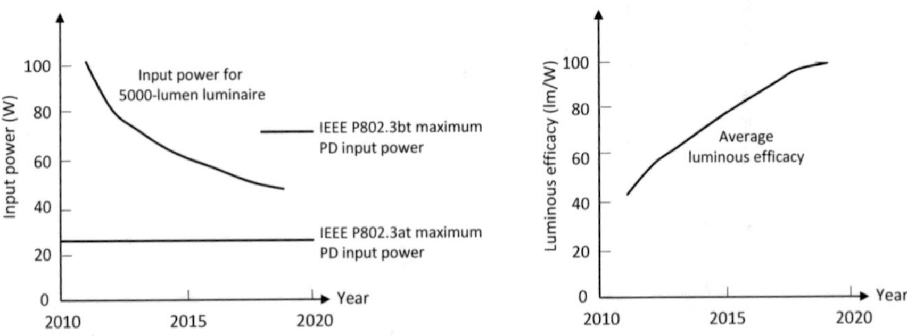

Fig. 3.9 LED power requirements and efficacy improvements in recent years

Table 3.2 Comparison of lighting systems

	Incandescent	LED
Light as percentage of power	10% Light	90% Light
Heat as percentage of power	90% Heat	10% Heat
CRI	100%	90+%
Envelope (air-tight glass enclosure)	Large	Small
Life expectancy	Short	Long
Maintenance cost/effort	Medium-to-high	Low
Color range	Limited color range: 2200–2900 °K	Extensive color range: 1200–10,000 °K

ongoing research for the development of white OLED devices for use in lighting applications. microLED is an evolving display technology where the displays consist of arrays of microscopic LEDs supporting the individual pixel elements—microLED affords energy efficiency, improved contrast and brightness, and faster response times; lighting applications are also being explored. Coating LEDs with phosphors to create a whiter color is a well-established technique; traditional approaches, however, suffer from a situation where a high percentage of the emitted light is reflected back toward the chip, thus not contributing to illumination. Active manufacturing research is underway to overcome this and other problems and seek to achieve higher efficiency goals. One approach involves physically separating the phosphor from the LED—rather than directly coating the surface of the LED—to reduce absorption (these systems are known as remote-phosphor systems) [35].

3.4.2 *Lighting Controls*

Lighting controls make lights "smart". Smart lighting entails the use of sensors to acquire parameters about the environment and the use of controls (including actuators) to implement parameter-driven decisions about the light output. There are "basic controls" and "more sophisticated controls". Sensors can be external from the fixture or can be built-in, the latter case being of the "In-Fixture Sensor Technology" type. Occupancy sensors are utilized to turn lights off and on. Ambient light sensors are utilized to dynamically control the luminance level of the LED lights in such a manner that a constant ambient light level is achieved throughout the entire day, while considering the brightness of the natural light that may be dispersed through windows (aka "Daylight Harvesting"). See Fig. 3.10. Control is achieved utilizing devices known as *controllers* that are embedded in the lighting system. In wired systems, dedicated (centralized) hardware-based controllers are placed between the sensors and the light sources they control. The controlling logic that implements some action when a specified value of sensor data is received, resides in the controller. In wireless (mesh) networks, the controller is a software component that is an intrinsic element of the device under control; sensors communicate directly

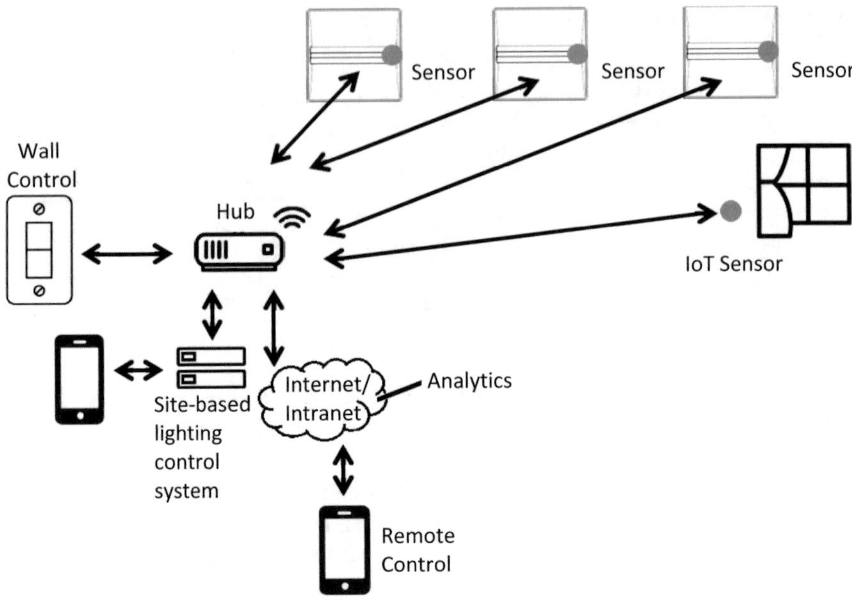

Fig. 3.10 Pictorial view of smart/IoT-based lighting system

to the soft(ware) controller embedded in the light or other elements (e.g., wall switch). See Fig. 3.11, that illustrates a comparison between dedicated (centralized) hardware-based controllers and distributed software-based controllers.

More broadly, typical control features include individual lights status, sensor status, space occupancy, estimated fixture lifetime (in support of maintenance and space management); daylight harvesting, occupancy control, time schedule, local switch/dimmer (in support of energy efficiency); dynamic white control, circadian cycle, mood enhancement, task-specific scenes (in support of comfort and productivity); integration with BMS, indicator/egress control, air quality monitoring, temperature and humidity (in support of broader integration). Occupancy sensors save up to 90% of lighting energy and daylight sensors save up to 25% of lighting (they can also be integrated with window shade control systems to maximize daylight harvesting while reducing glare) [36]. Lighting controls are now included in all key building energy codes, such as IES/ANSI/ American Society of Heating, Refrigerating and Air-Conditioning Engineers (ASHRAE) Standard 90.1-2016 which applies to major space renovation projects and in all new building construction. Current recommendations (e.g., by the International Code Council) dictate that occupancy sensors must be installed in: conference rooms, copy rooms, employee break rooms, employee lunch rooms, enclosed offices, janitor closets, locker rooms, lounges, open plan office areas, private offices, restrooms, storage rooms, training rooms, warehouses, and all other spaces 300 square feet and less enclosed by floor to ceiling partitions.

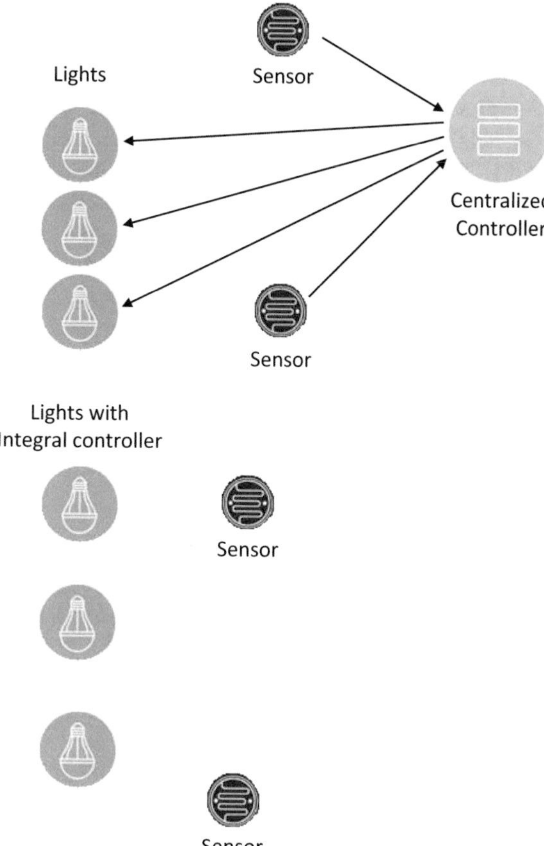

Fig. 3.11 Centralized vs distributed controllers

Initially the sensors were stand-alone, for example to establish the presence of people in the proximity of the light source. More recently, the concept of a set of connected sensors (over a wired or wireless network) has emerged. Some organizations are in fact contemplating a new form of network convergence: they envision connecting and possibly also powering extant stand-alone building systems (e.g., Heating Ventilation and Air Conditioning [HVAC], lights, security cameras, card readers, elevators) over an Ethernet/IP network to endeavor to improve efficiency, automate processes, and generate analytical insight that can be used for actionable service improvement initiatives.

Traditionally lighting control systems employed analog interfaces or proprietary digital control methods. Both approaches have limitations. Hard-wired analog (dimming) systems are complex, inflexible, and expensive to implement and cannot control individual lights. While providing increased flexibility over analog systems, proprietary digital controls typically do not allow one to control individual ballasts and are relatively complex and expensive to deploy and maintain. Some newer systems address and mitigate these issues.

Analog controls require point-to-point wiring for each fixture in any control zone and dedicated occupancy and/or daylight sensors for each zone; additionally, it limits the number of drivers per switch/dimmer and 15A circuit. Traditional examples include (also see Fig. 3.8) Analog 2-wire+G (Forward Phase), Analog 2-wire+G (Reverse Phase), Analog 3-wire+G; and Analog Low Voltage (LV) 0–10 V (aka 1–10 V and 10–0 V). Digital controls have no specific wiring requirements to assign control zones; each driver/fixture is addressed individually; one can achieve full power load on a 120 V/277 V 15A circuit—no driver limits per switch/dimmer. Traditional (wired) examples include Digital DALI, Digital EcoSystem, and Digital DMX [36]. In the commercial arena, about 25% of the installed systems had sophisticated lighting control systems at press time with the penetration expected to grow 1% a year, to a level of 40% by 2035 [33].

3.4.3 Lighting Controls Connectivity

Controls can use wireless or wired links. Wireless systems in use include, among others: ZLL, Bluetooth Mesh Network, Wi-Fi, Li-Fi, LoRa/Sigfox, and NB-IoT/LTE-M. Bluetooth and ZigBee currently dominate the wireless IoT connectivity space. Wired systems are covered in Sect. 3.4.3.7.

3.4.3.1 ZigBee

ZigBee is expected to capture a major share of the smart lighting markets in the near future. ZLL is a standard for interoperable (consumer) lighting and control systems. Although it can operate outdoors the principal application is for indoor applications (the use of ZigBee for outdoor applications is somewhat challenging because signal strength depends on uncontrolled outdoor conditions; there are also challenges in the presence of heavy Wi-Fi traffic [37, 38]). ZLL operates at 2.4 GHz, supporting worldwide operation; it has a range up to 70 m indoors. Version 1.10 of the standard was issued in 2012 by the ZigBee Alliance and was developed by firms such as Philips, Sylvania, GE, and Texas Instruments, among others. ZLL enables users to gain wireless control over the LED fixtures and bulbs, switches, timers, and remotes; users are able to change lighting from remote locations to adjust to the ambiance, season, or occupant's specific function or task [3, 39]. ZLL is an application profile (among several others), typically running over ZigBee PRO at the network layer and on IEEE 802.15.4 Media Access Control/Physical layer (MAC/PHY) (manufacturer-specific extensions can be added).

Touchlink (more specifically the Touchlink Commissioning Protocol) is a simple commissioning mechanism positioned for the user, such as a consumer—it eliminates the need for a coordinator. The protocol employs inter-PAN (Personal Area Network) communication to transfer commissioning messages (e.g., scan request/response, device information request/response, identify request, start/join network request). Devices in a given environment discover each other via the Touchlink commissioning process. ZLL and ZigBee Home Automation (for example to control thermostats, shades, and security appliances) can be used in combination. ZLL can also be used via an Internet gateway (router) to support remote access. Security extensions to the protocol have been sought in recent years, since there are documented security defects of ZLL Touchlink Commissioning Protocol [40–42]. ZLL does not utilize a coordinator or trust entity, and consequently the traditional ZigBee security mechanisms are not usable. The original security mechanism is as follows: ZLL makes use of network level security where both sides (user, device) exchange a network encryption key; the Touchlink initiator is responsible for generating the key and for transmitting it to the target device during the commissioning phase. To ascertain that the key is not transmitted "in the clear", the key is encrypted with a ZigBee Light Link master key—this encryption key is assigned by the manufacturer once a device has successfully completed ZigBee certification. The initiator generates a random local encryption key and encrypts it using the master encryption key; the initiator requests a targeted device to either start a network or join the initiator's network, delivering the encrypted key at the same time. The target device decrypts the key using the master key after which normal communication is now secured with the local key.

3.4.3.2 Bluetooth

Recent advancements in Bluetooth® technology facilitate the deployment of scalable, smart building applications in commercial and industrial environments, with an emphasis on reliability, performance, security and multi-vendor interoperability. The specifications describing Bluetooth mesh networking were published in 2017, specifically V1.0. This new Bluetooth capability is designed for applications such as smart buildings, smart factory, commercial lighting, and smart industry, including connected lighting, building automation, and sensors. Bluetooth mesh expands the capabilities of Bluetooth, complementing other Bluetooth systems such as Bluetooth Low Energy (LE) and Bluetooth Basic Rate/Enhanced Data Rate (BR/EDR), each of which have intrinsic strengths and application foci.

- Bluetooth BR/EDR was developed for the transmission of a predictable or isochronous stream of data over point-to-point connections between two devices;
- Bluetooth LE was designed for power-efficient point-to-point communication as well as broadcast data in a one-to-many environments.

Bluetooth mesh networking is not per se a radio transmission technology, but a connectivity technology that enables many-to-many networks of large groups of Bluetooth devices to be established. Messages generated by one node can transfer from node to node over the network until these messages reach their destination, allowing communication to take place significantly beyond the direct radio range of each individual device. Copies of messages can transit via multiple paths in the network, achieving transmission redundancy and accruing the benefit of high reliability to the network, without the need to predefine special rules or routes. Multi-path and multi-hop delivery are intrinsic in the way Bluetooth mesh works [43]. Bluetooth mesh makes use of Bluetooth LE (version 4.0 or better) for wireless radio communications; in the protocol stack, Bluetooth LE (modulation scheme/air interface) is positioned underneath the Bluetooth mesh functionality (LE provides the radio interface fields, such as a preamble, an access address, and a CRC). There can be 32,767 elements in a mesh network, and 127 Hops can be traversed by a mesh message. Bluetooth LE 4.x supports a symbol rate of 1 mega-symbols per second (Msps)—thus, Bluetooth LE 4.x is faster (four times faster to be exact) than other mesh technologies, for example based on IEEE 802.15.4, which runs at 0.250 Msps. Typically mesh messages can fit in a single mesh network PDU.

Bluetooth mesh networking facilitates the coverage of large environments; is optimized for low energy consumption; and makes efficient use of radio resources, supporting scalability. In the view of proponents, BLE provides companies with relevant use cases, a sophisticated solution, and an optimal cost-benefit ratio. Bluetooth hardware (beacons) has been on the market for several years and has most often been used in the context of proximity applications [3, 44]. The Bluetooth® SIG (https://www.bluetooth.com) anticipates 360,000,000 annual Bluetooth smart building device shipments by 2022, and 1.3 billion sensor deployments for condition monitoring in smart buildings by 2020.

An important consideration is total capacity given than any group of devices within radio range and using the same frequency (set) are sharing the limits of the available (spectrum) resources (but not those further away in view of the concept of space division multiplexing). The Bluetooth spectrum is a shared medium and collisions are one of the major considerations. For Bluetooth mesh networking, the ultimate question depends on the application. For example, for a mesh network consisting of streetlights, there is typically a regular uniform distribution of well spaced-out nodes. Thus it is unlikely that a particular node is within distance range of many other nodes; it follows that there is little contention of the shared radio spectrum. However, on the floor of a commercial building, with a dense deployment of nodes, the majority of mesh nodes are typically all within direct radio range of each other, thus competing for the same radio resource.

As long understood from the extensive work on random access techniques, in a wireless environment, packet collisions degrade the overall transmission performance. To address this issue, Bluetooth mesh Protocol Data Units (PDUs) have been defined to be small: a shorter PDU leads to fewer collisions—Bluetooth mesh PDUs are of *at most* 29 bytes long—much shorter compared to the PDU size of other mesh technologies. In fact, generally, 11 bytes for the application payload has been determined to be sufficient, along with 1–2 bytes for an opcode and up to 10 bytes for parameters, for example a value measured by a sensor, or a multidimensional light value such as luminance/brightness, hue, saturation, and a transition time. For example, a typical message to switch devices on or off, is only 22 bytes in length. There will be some additional encapsulation data in the underlying Bluetooth LE packet (18 bytes) [45]. In summary, the small PDU

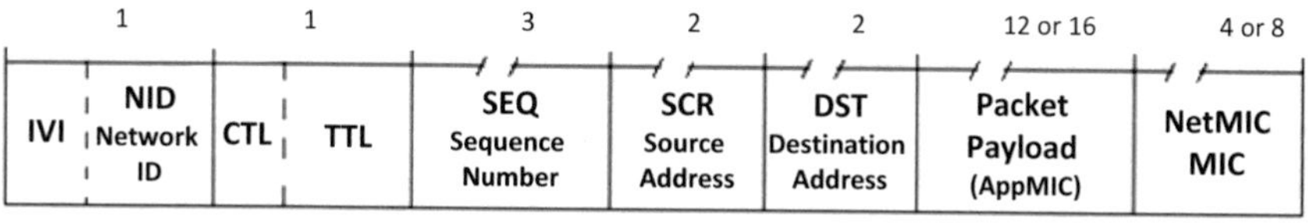

1		1		3	2	2	12 or 16	4 or 8
IVI	NID Network ID	CTL	TTL	SEQ Sequence Number	SCR Source Address	DST Destination Address	Packet Payload (AppMIC)	NetMIC MIC

Fig. 3.12 Bluetooth mesh networking PDU. *IVI* initialization vector index, *CTL* network control message indication, *TTL* time to live, *SRC* source, *DST* destination, *MIC* message integrity check

size of Bluetooth mesh and the high symbol rate of the Bluetooth LE radio both decrease the required transmit time for a PDU, thus optimizing performance. With the LE overhead the message is 47 bytes; a single transmission of a Bluetooth mesh message requires only 400 μs; this an order of magnitude less than when using other wireless technologies, and with the emerging 2 M Bluetooth 5 PHY the performance is further improved.

BT 4.x specification has a 33-byte packet format. The Bluetooth mesh packet size is shown in Fig. 3.12; the packet has a 12 or 16-byte payload. For larger payloads, there is a process of segmentation and reassembly. Bluetooth mesh enjoys a higher data rate but the packet payload is smaller; therefore, it requires a higher number of packets to send the same amount of data. In Fig. 3.12, if the Message Integrity Check (MIC) is limited to 4 bytes, the total PDU length is 29 bytes.

In addition to the payload there are "overhead" bytes to deal with control and security, specifically addressing/propagation control (5 bytes: SRC, DST, CTL+TTL) and security (8 bytes: IVI+NID, SEQ, AppMIC / NetMIC). TTL is a parameter utilized to limit the number of hops that a message can travel as it is being relayed (a value of zero implies that the message is intended to be received only by nodes that are in direct radio range of the transmitting node). The 1-byte IVI+NID aids in identifying a network; the 3-byte SEQ in conjunction with the slowly-propagated IV Index, comprises a 7-byte sequence number—each PDU transmitted over on a mesh network has a unique sequence number, per given SRC address. The novelty is the inclusion of just 3 bytes in the air interface PDU; the other 4 bytes are slow changing and are known to the network. The sequence number is important for detecting replayed packets.

Specific mesh messages are utilized to make optimal use of the underlying Bluetooth LE resources, as described in the bearer layer of the mesh protocol stack. Two bearer mechanisms are defined; of these, the *advertising bearer* is utilized to enable communication between nodes in the mesh network. This bearer defines how PDUs are to be broadcast within Bluetooth LE advertising packets on the three Bluetooth LE advertising channels and how they can be received by nodes scanning for PDUs on these channels. Generally, identical copies of each PDU are sent on (up to) three advertising channels to increase transmission reliability.

Bluetooth Mesh incorporates both network layer security and application layer security. Messages may be secured with two independent encryption keys. This mechanism allows for relay nodes to authenticate a message on a network layer while precluding tampering with the application payload (for example a light bulb that relays a message to a door lock cannot change the payload from *open to close*, but it can validate if the PDU belongs to its own network). MICs determine the level of security of the system. The network layer MIC (NetMIC) as well as the application layer MIC (AppMIC) can be either 4 or 8 bytes long. Bluetooth Mesh developers thus perceive a level of security that is claimed to be "*sufficient for almost any building automation, lighting control, and sensor application*" [45].

Typically, a mesh network needs to support a physical space that is larger than the immediate radio range of individual nodes. Bluetooth mesh allows designated nodes to be configured to act as *relay nodes*, retransmitting every message they receive. Relay nodes are often deployed on the edge of building quadrants or zones to extend the range of a set of devices and enabling messages to reach another set of devices in that other zone. Relays can be used to ensure reliable network operation; however, the retransmission of messages utilizes spectrum resources, thus, they should be deployed sparingly and in meaningful locations on the commercial floor under consideration.

Furthermore, there are mechanisms to reduce traffic by being able transmit messages to a group of devices. Bluetooth mesh allows a message to be sent to a specific device utilizing a unicast address and allows a message to be sent to a set of devices using a group address. The system utilizes a publish-and-subscribe messaging mechanism such that messages are processed by every device subscribed to the group. For example, to dim lights on a floor (say in the middle of the day) one needs only a single message to be transmitted, not a new message to each target device. Group addressing is the typical mode of operation; unicast addressing is infrequently employed (e.g., used for configuring new devices when they are added to the network). In a Bluetooth mesh network, the controller is a software component in the Light Lightness Control (LC) Server Model, which is part of the overall software embedded on the light module itself. The control logic is an intrinsic element of the device under control and sensors are able to communicate directly with the control logic by publishing mesh sensor messages, to which, in turn, the light's Light LC Server subscribes. This decentralized, software-based approach with group messaging is an efficient construct, with fewer messages needed for specific operation type compared with the centralized approach using dedicated controller devices and unicast messaging. Bluetooth mesh makes use of both *acknowledged messages and unacknowledged messages*. Acknowledged messages need a response from nodes that receive and process them; unacknowledged messages do not require a response. Request-response mechanisms do not scale effectively in cases where the majority of interactions are multicast in nature, for example in the Bluetooth mesh environment. Thus, the use of request-response is kept to a minimum. Figure 3.13 depicts one possible physical architecture of a Bluetooth-based LED lighting system.

3.4.3.3 Wi-Fi

From a general perspective, Wi-Fi technology is one of the most common and well-entrenched local wireless technologies, having been around for decades. Despite

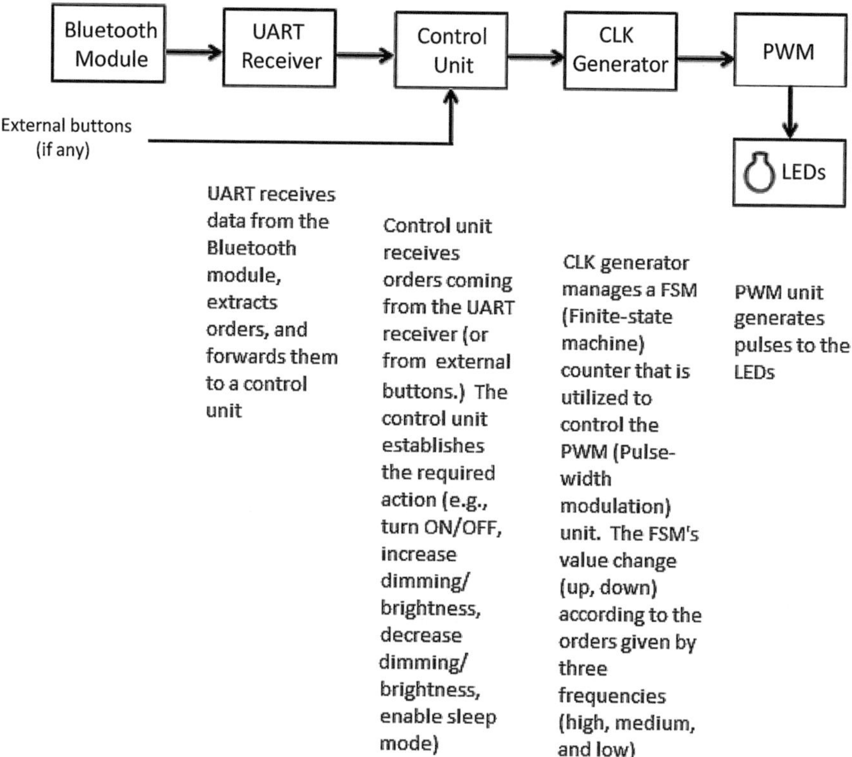

Fig. 3.13 Example of Bluetooth-enabled LED lighting

its high energy consumption, Wi-Fi is perceived as one of the main wireless communication solutions for the majority of Industrial IoT (IIoT) devices. However, many generic IoT sensors—including lighting sensors—do not (yet) implement Wi-Fi on a broad scale. Some recent Wi-Fi standardization efforts, mainly focused on industrial applications, include 802.11af (White-Fi) and 802.11ah (HaLow). To address the channel saturation of already-deployed 2.4 GHz Wi-Fi systems and avoid the cost of cellular licensed spectrum, IoT device manufacturers are assessing the use of new sub-1GHz Wi-Fi bands—however, the achievable bandwidth in the sub-1GHz spectrum is limited. In addition, new bands in the 3.5 GHz area of the spectrum for the Citizen Broadband Radio Service (CBRS) are being explored in the U.S. Furthermore, there are also standards for Wi-Fi systems using millimeter waves.

Up to the present time, lighting applications have tended to use other wireless technologies than Wi-Fi; however, some Wi-Fi implementations have evolved. In addition to stand-alone lighting applications (where the LED fixture incorporates a Wi-Fi transceiver), an evolving application that employs Ethernet technologies entails PoE-enabled lighting [46]. An Ethernet Gateway provides a multipurpose

Ethernet connection to a lighting control system. Certain types of bulbs now on the market do not require a bridge or hub to connect over a Wi-Fi network. However, Wi-Fi RF signals are difficult to confine to a defined space and can be intercepted externally to a building. Wi-Fi has a limited set of usable frequencies, and although newer IEEE 802.11 standards (for example, 802.11ac and 802.11ax) allow higher throughput, there is the risk of congestion with the plethora of devices, laptops, PCs, and smartphones using Wi-Fi in a given space. Lighting is a critical (safety) systems that needs to be properly safeguarded. Corporate Information Technology (IT) management may not be receptive to running the lights over their network and this opinion is also shared by some of their Facility Department counterparts.

3.4.3.4 Li-Fi

Li-Fi is an optical communication technology that utilizes light to transmit information between devices, including visible light, ultraviolet light, or infrared signals; LED sources are employed to generate the information-carrying beam. It is an example of a VLC. The spectrum utilized by Li-Fi channels can be significantly larger than the radio frequency spectrum used by Wi-Fi systems, resulting in high throughput where data rates in the multiple Gbps can be achieved in some systems. Li-Fi transmission is impervious to traditional electromagnetic interference thus it can also be used in specific environments such as hospitals. The light switching occurs at a very high speed thus it cannot be noticed by the human eye; also, the light can be dimmed to below human visibility while still emitting enough light to convey information. Li-Fi enjoys smaller round-trip latency than Wi-Fi. A Li-Fi Access Point (AP) can be used for WLAN-like applications to user's workstations and laptops. Additionally, sensors (whether in the fixture or in the floor environment) can use VLC to communicate information to the lighting control system. The light waves cannot penetrate walls thus the technology is limited to a short range, although full direct line of sight is not necessary since the light is typically reflected off the walls. Applications include, among others, home and building automation, hospital medical systems, vehicles, and industrial automation. While the IEEE 802.15.7 standard does not per se consider the use of optical orthogonal frequency-division multiplexing (O-OFDM) modulation methods, the standard defines the PHY and the MAC layer. The standard defines three PHY layers:

- The PHY 1 is designed for outdoor environments and supports data rates up to 267 kbps using on-off keying Manchester coding and variable pulse position modulation (VPPM);
- The PHY 2 supports data rates up to 96 Mbps also using on-off keying Manchester coding and VPPM;
- The PHY 3 supports rates up to 96 Mbps utilizing color shift keying (CSK).

VLC smartphone prototypes were demonstrated in 2014. Philips Lighting now Signify has developed a VLC system for stores.

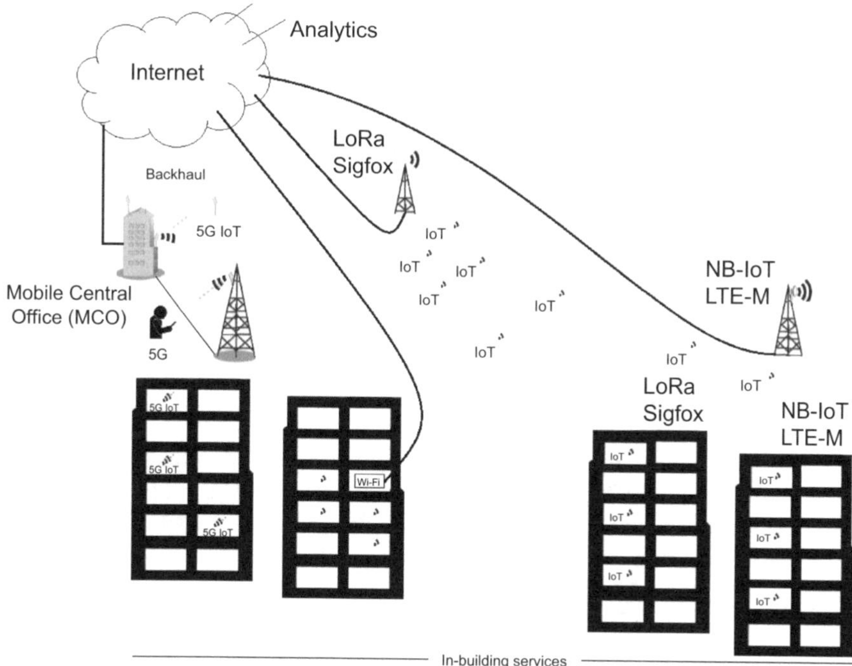

Fig. 3.14 LoRA, Sigfox, NB-IoT, and LTE-M examples

3.4.3.5 LoRa and Sigfox

LoRa and Sigfox are examples of LPWAN (Low Power Wide Area Network) technologies. The transmission can reach several miles to a centralized antenna connected to the (Internet) cloud. LoRa can be used over a distance of 6–15 miles with Line of Sight (LOS); Sigfox can be used for 30 miles in rural environments and 1–6 miles in city environments. Both technologies operate below 1 GHz. They are proprietary systems that have received some penetration in the IoT environment. See Fig. 3.14.

3.4.3.6 NB-IoT and LTE-M

NB-IoT (also known as LTE Cat-NB1) is a licensed low power LPWAN technology providing cellular-based connectivity for IoT devices, that can coexist with existing LTE specifications. NB-IoT was advanced by 3GPP in its LTE Release 13. NB-IoT has received support from Qualcomm, Ericsson, and Huawei, among numerous other vendors and service providers. NB-IoT uses Direct Sequence Spread Spectrum (DSSS) modulation within a 200 kHz cellular channel.

LTE-M (Long-Term Evolution Machine Type Communications) Rel. 13 (Cat M1/Cat M) is also a licensed low power LPWAN technology based on the LTE cellular network architecture; as such it is LTE compatible, easy to deploy, and does not require the deployment of new cellular antennas. It uses 4G-LTE licensed spectrum bands below 1 GHz and is a power-efficient system. LTE-M allows the upload of 10 bytes of data a day but can also allow access to rates in the Mbps range—therefore, it can support several use cases. *Major* U.S. carriers including AT&T and Verizon offer LTE-M services (as noted, Verizon has also announced support for NB-IoT, while T-Mobile/Sprint appears to lean in the NB-IoT direction). See Fig. 3.14.

3.4.3.7 Wired Systems

Wired systems include, among others, Power-line Communication (PLC), Electronic Control Gear (ECG), RS-485, DMX, DALI, KNX, and LON (see Fig. 3.15). Note a newer wired method for lighting is PoE discussed separately in Sect. 3.4.5.

A brief description of these systems follows.

- A number of PLC approaches are technically feasible. However, as yet none of these systems and protocols have reached the level of market penetration and cost competitiveness, although there are activities underway by a number of industry and standards-organizations to develop standards, particularly for Smart Grid applications—for example, the European Commission (EC) M/411 Smart Metering Mandate and EC's M/490 Smart Grid Mandate: EC Mandate.

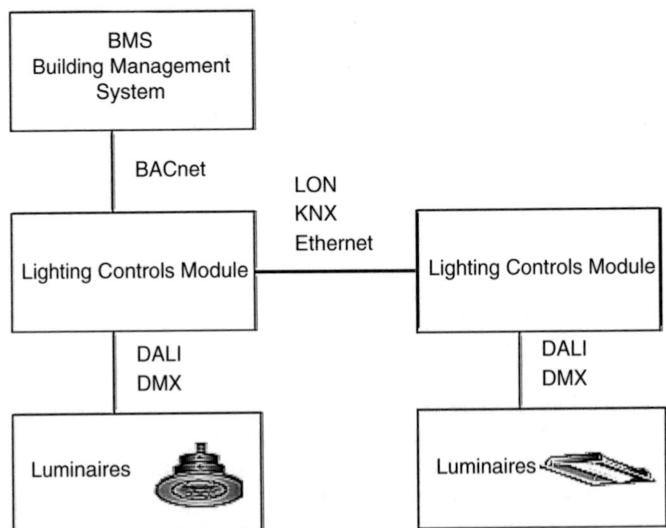

Fig. 3.15 Examples of wired control (in a BMS environment)

- ECG is widely used in low-complexity lighting systems. Control gear regulates the output of light from the luminaire by using analog 1 V–10 V signals. The dimmer's setting is transmitted via a separate control line. This system is inflexible because its allocation cannot be changed: the grouping of the luminaires is determined by the circuits in the initial installation and any change requires a new arrangement of the connection and control lines.
- RS-485 has long been used in Supervisory control and data acquisition (SCADA) applications.
- The DMX digital control protocol is principally used for stage lighting (e.g., media facades or stage-like room lighting effects). Each luminaire has a bus address. The control information is transmitted uni-directionally over a dedicated 5-core cable at a transfer rate of 250 kbps (the DMX 512-A is a more recent version that allows for bi-directional transmission) [19].
- DALI is an open industry standard (IEC 62386) allowing DALI-compliant components (e.g., control systems, sensors, ballasts, controllers, switches) from different manufacturers to operate seamlessly as a complete system. DALI is promulgated by the Digital Illumination Interface Alliance (DiiA®) and the terms DALI and DALI-2 are trademarks owned by DiiA. DALI is a control protocol that enables the control of luminaires that have individual DALI control elements. DALI supports data communication between the individual components of a lighting system. In recent years, it has become well known as a commercial lighting control protocol that works with KNX; DALI is being used in commercial applications such as offices due to its flexibility and ease of installation, but has not been used residentially due to the relatively high cost and poor product availability of DALI-ready luminaires [47]. A DALI system can be designed as a stand-alone system or a sub-system to a BMS, in which case it communicates bi-directionally through a gateway. The two-wire 1.2 kbps control line can be run together with the mains supply cable in a 5-core cable and the bidirectional channel allows feedback from the luminaires on different aspects such as lamp operation and/or failure [48]. Each DALI loop can individually address and control up to 64 devices; broadcast (group) addressing is also supported. Ballasts are connected to a DALI controller utilizing standard building wire; various DALI controllers can be connected to each other for unified control of large office/building areas. A DALI system allows the creation of up to 16 groups of ballasts, and individual ballasts can be assigned to a specific group or to all of the groups. Each group can also have up to 16 settings for various lighting scenes [49]. Bi-directional communication allows ballasts to provide feedback information such as: lamp condition, lamp energy level, on/off luminaire state, and ballast condition. DALI-2 added the standardization of control devices such as daylight sensors, room occupancy sensors, manual lighting controls (e.g., push-buttons, sliders), and application controllers (these being the so-called "brains of the system"). DALI-2 has been proposed as the global standard for smart, digital, lighting controls in the IoT era. Additionally, the DiiA published a number of new specifications in 2018 and 2019, extending DALI-2 functionality with power and data, particularly for intra-luminaire DALI systems. Applications

include indoor and outdoor luminaires, and small DALI systems. IEC 62386-209 describes color control gear (added to DALI-2 certification in early 2020); IEC 62386-202 describes self-contained emergency lighting including automated triggering of tests; and, IEC 62386-104 describes wireless and wired transport alternatives to the conventional wired DALI bus system. The D4i trademark is used on certified products that incorporate these new features.

- KNX is a standardized digital control system that controls lighting and other building automation systems such as HVACs and solar screening equipment. Sensors and actuators have individual addresses. The data is transmitted bi-directionally over a separate 24 V control line unshielded twisted pair (UTP) cable at a rate of 9.6 kbps; a prioritization scheme prevents data collisions. KNX can be utilized in large installations.
- LON is a standardized digital control protocol that controls building systems; it is also used in industrial and process automation. It is TCP/ IP compatible. The data communication takes place over a UTP cable (as a dedicated control line) and operates up to 1.25 Mbps.

3.4.4 Building Management Systems: Possibly Supporting Lighting

A BMS is a multi-system platform that is utilized to monitor and control a building's electrical and mechanical equipment, such as electrical systems, HVAC, elevators, plumbing, security/surveillance, and alarms; some BMS systems also provide integrated control of lights. A BMS enables facilities managers to optimally control energy management: the BMS software interacts with controls elements in the various mechanical/electrical systems in the building to monitor and regulate the energy used in real-time. BMSs are often used to implement Demand Response (DR) arrangements. Typically, BMSs can be accessed remotely to manage loads and enhance building efficiency, thus allowing organizations to reduce the energy required to heat, cool, ventilate and illuminate, a building. IoT principles are expected to enhance, standardize, and extend the service function and the service scope of BMSs.

In the recent past one has observed an advantageous convergence of a multitude of building-supporting technologies and systems to a TCP/IP-based infrastructure implemented via an intranet (in multi-tenant buildings a building-oriented intranet may be required). Initially various building systems utilized individual protocols and cabling systems; this predicament is inefficient from both a deployment perspective as well from a system administration perspective.

BACnet is an early effort at some standardization: it is an ANSI, ASHRAE, and International Organization for Standardization (ISO) standard communications protocol (ISO 16484-5) for building automation and control. It defines several services that are used to communicate between control devices found in buildings (including HVAC, access control systems, fire detection systems, and lighting control systems and it specifies a number of network-, data link-, and physical-layer protocols. BACnet developers realized that it would be easier to install a common cabling infrastructure (for example, UTP Category 6 or 6a cable) for all building-related functions, and they also realized that it would be easier to migrate to a common set of upper layer protocols (e.g., the TCP/IP suite). Although BMSs have migrated to IP-networking (which has the additional benefit of allowing several buildings to be remotely monitored by a centralized operations center, say using cloud-based analytics), IoT will bring these capabilities to the next level; in fact, BMSs are now often based on IoT principles. Intelligent lighting controls and HVAC optimization are just two of the use cases that are facilitated by the IoT—as noted earlier, smart lighting not only allows intelligent centralized (and/or remote) control but also lowers energy consumption while improving the occupants' experience. An IoT-ready BMS can be utilized to manage many other building functions such as access, fire detection, surveillance, and so on.

3.4.5 PoE Technology

PoE aims at delivering DC power over Ethernet cabling, thus simultaneously supporting data transmission and power. IEEE has defined three PoE standards over the years: the original IEEE 802.3af-2003 PoE standard; the 802.3at-2009 PoE standard (also known as PoE+ or PoE plus); and, the IEEE 802.3bt-2018 (also known as 4PPoE). IEEE 802.af has been positioned and used for VoIP phones, IP cameras, and wireless Access Points; the newer standards are being positioned for the "digital building", which also encompasses PoE/LED lighting. PoE is supported in 10/100/1000BASE-T/TX networks. 802.3bt adds Type 3 and Type 4 Power Sourcing Equipment (PSE), new Classes 5-8, with the ability to provide higher power—45 W, 60 W, 75 W and 90 W—and extends PoE to 10GBase-T. In summary, PoE spans the IEEE 802.3 af/at/bt standards.

PSE are devices capable of providing power over and Ethernet cable; Powered Device (PDs) are endpoint devices capable of receiving power over an Ethernet cable from a PSE. PSEs are power supplies designed to provide power to a PD. A PSE is also typically a PoE Switch. PDs are endpoint devices that accept and utilize power from the Ethernet cable for their operation. Additionally, the IEEE standard defines two power configurations. In the first configuration, power is supplied via PSE located in the switch/hub (endspan). In the second configuration, power is provided via insertion in the network using a PSE located in a midspan; this configuration allows for support of legacy elements and enhanced control of which network segments are powered. Midspan (also known as a Power Injector) is a type of PSE that is a pass-through RJ45 element that adds PoE power. Furthermore, there are several power-related configurations: Type 1 PSE (or PD) is an 802.3af compliant PSE (or PD); Type 2 PSE (or PD) is an 802.3at (PoE+) compliant PSE (or PD); Type 3/4 PSE (or PD) is an 802.3bt compliant PSE (or PD). For example,

- Type 1 (over two pairs) (older legacy installations): the PSE is able to supply a maximum of 15.4 W over a voltage range of 44–57 VDC making use of Category 3 cabling or better.
- Type 2: the PSE is able to supply 30 W (over two pairs) or 60 W (over four pairs) over a voltage range of 50–57 VDC making use of Category 5 or better cabling.

Type 3 and Type 4 PSE are included in 802.3bt—See Table 3.3. Clause 33 of IEEE Standard 802.3-2015 specifies the use of twisted-pair Ethernet cables for PoE applications, with each cable being composed of eight conductors (i.e., four pairs) and being terminated with RJ45 connectors. IEEE 802.3 specifies the use of Category 5 or better quality cables, limits the overall resistance of the link section between PSE and PD, and limits link section length to 100 m; the standard points to Telecommunications Industry Association (TIA) 568-C.2 performance requirements for Category 5e or better quality cabling. The 802.3bt Type 4 PoE++ standard mandates a different type and use of the available Ethernet cable and connector wiring, while maintaining compatibility with the traditional use and data-link requirements.

Classification is the process where the PSE establishes the PD's power requirements. The PD negotiates a power class during the initial activation of the connection—the PSE must determine the power classification of the PD, effectively how much power must be supplied to the PD. In IEEE 802.3at environments the PD is limited to a maximum of 13.0 W for Type 1, 25.5 W for Type 2 over two pairs (51 W for Type 2 over four pairs)—the voltage range between 37 V to 57 VDC. For office building lighting applications, 802.3bt Type 3 or 4 are the most practical standard. The typical voltage losses in the Ethernet cables and connectors are commonly referred to as I^2R losses, because the power dissipated by a conductor is the product of the square of the current (I) it carries and its resistance (R).

There are two mechanisms for determining the PD's power requirement: physical layer classification and data link-layer classification.

- Physical Layer Classification utilizes the cable and PD electrical characteristics to determine which power class to assign the PD (PD Type 1 or PD Type 2). For example, in the IEEE 802.3at environment, when the PSE detects a connected PD, it sends a pulse to establish how much power the PD requires; based on how much current is drawn during this pulse, the PSE classifies the PD into one of several classes, specifically class 0–4 (or class 0–3 for Type 1 PSEs and PDs).
- Data Link-Layer Classification, where after the data link is established, the PD and PSE communicate utilizing the Link Layer Discovery Protocol (LLDP). This classification method has more granular power resolution (1.11 W increments) than is the case with the physical layer classification and supports dynamic power allocation during PD operation, with the PD periodically communicating its power requirements to the PSE.

Table 3.3 Various PoE standards and some key parameters

Standards	Class	Supported cabling	Supported modes	PSE output power (W)/PD input power (W)	Power management	PD Type
802.3af 802.3at type 1, PoE	1	Category 3 and Category 5	Mode A (endspan), Mode B (midspan)	4/3.8	Power class levels 1–3, negotiated by signature	1
	2			7/6.5		1
	3/0			15.4/13		1
802.3at type 2, "PoE+", "PoE plus"	4	Category 5	Mode A, Mode B	30/25.5	Power class levels 1–4 negotiated by signature or by LLDP	2
802.3bt type 3, "4PPoE", "4P PoE", "4-pair PoE", "PoE++" sometimes "UPOE", a Cisco system	5	Category 5	Mode A, Mode B, 4-pair mode	45/40	Type 3: Power class levels 1–6 negotiated by signature by LLDP	3
	6			60/51		3
802.3bt type 4, "higher power PoE"	7	Category 5	4-pair mode	75/62	Type 4: Power class levels 1–8 negotiated by signature by LLDP	4
	8			90/73		4

Per the IEEE 802.3at standard, if a Type 2 PSE classifies the connected PD into Class 4 (PoE+ power), a second handshake is needed (initially the power is restricted to the IEEE 802.3af limits). This second handshake allows the PD to establish that full 802.3at-defined power is available, and, for safety reasons, it also allows the PSE to validate that a Type 2 PD is in fact connected to the port. This mechanism has two implementations: a hardware-based Two-Event Classification or a protocol (software)-based LLDP classification.

- In the hardware-based Two-Event (2-Event) classification, the initial voltage pulse classification event is repeated and once the PD passes the second classification event as a class 4 PD, it will have full class 4 802.3at power.
- In the LLDP classification the 802.3at devices are TLV (type-length-value or tag-length-value) enabled, allowing the PSE and PD to negotiate power requirements.

Typically, the Two-Event classification allows negotiation up to 25.5 W input power; LLDP allows negotiation up to 51 W input power. Note that making allocation of power based solely on classification is problematic since classification is only a coarse procedure for determining the actual power consumption—advanced PSEs can sense dynamically the amount of power actually required by the PD or utilize a combination of classification with dynamic sensing.

Some Layer 2 switch vendors (e.g., Cisco) add proprietary PoE features such as Fast PoE and Perpetual PoE to enhance the availability of port power after a hypothetical power failure. Perpetual PoE enables PDs connected on specified ports to continue to receive power during a soft reload/reboot of the switch (this feature is certainly important for PoE Connected Lighting). Fast PoE permits illumination with low brightness within 10 s of a hard reload (again, this feature is important for PoE Connected Lighting since switches can take several minutes to reboot).

As already noted, for office building lighting applications, 802.3bt Type 3 or 4 are the most practical PoE standards. As alluded to above PoE entails a series of handshaking protocols, to manage the power source (the PSE) and sinks (the PDs) for performance and capability. These mechanisms have become more complex with each new upgrade to the PoE standard. Therefore, IC manufacturers need to develop multifunction devices which address the PoE power-delivery and control issues in the context of the requisite protocols and timing. PoE is also being used for industrial control system applications including the support of sensors, controllers, meters, cameras, robots, and so on.

3.4.6 PoE Lighting

PoE can supply power to LED-based luminaires in lieu of traditional AC power. The higher wattage systems are germane to office or factory lighting applications. Figure 3.16 provides an example of a PoE lighting arrangement. First-generation PoE lighting systems experienced substantial voltage drops since each fixture had its own direct run to a distant layer 2 switch in some IT switch room; to

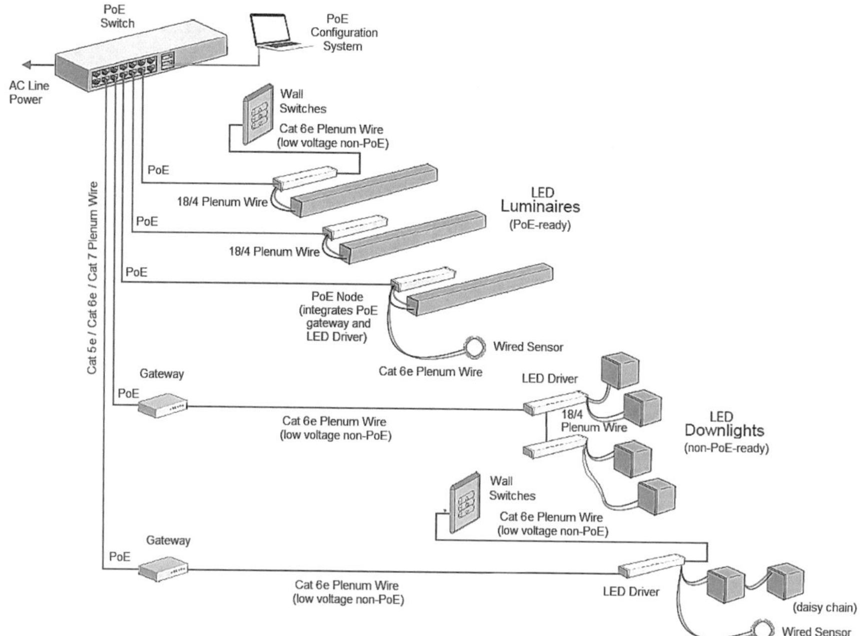

Fig. 3.16 Example of a PoE lighting arrangement

address this voltage loss issue a number of vendors have developed "distributed" switches that can be integrated in the plenum [50]. The ability to do luminaire daisy-chaining and the higher input power supported by the latest PoE standards mitigate some of these concerns. A possible advantage of PoE-based lighting is the ability to use LV-qualified installers, which might be, in theory and in some cases, less expensive that traditional electricians. Important consideration related to PoE lighting include cable bundling (over)heating, backup power, emergency lighting, cybersecurity, technology risk/obsolescence, and functional corporate ownership (more on this later). When the lighting system is connected using PoE technology it is straightforward and natural to make the light or luminaire "smart" and remotely controllable. However, since LEDs are current-driven devices, they require a voltage-to-current source converter (a LED driver) between the PoE rails and the LED fixtures.

Emergency (EM) lighting is often dictated by various local ordinances. The common practice for EM lighting is to utilize line-voltage power, either making use of fixtures with built-in batteries and transfer circuitry from a normal panel, or bringing power to designated fixtures from emergency lighting panels (some lighting control systems use a hybrid along with 0–10 V—or other—dimming signal. In the PoE Lighting environment some alternate approaches are needed; one approach is to use an uninterruptible power supply (UPS) connected to the Layer 2 switch.

3.4.7 Lighting Standards

There is a large number of standards and/or recommendations regarding (office) lighting. Key industry thought leaders include the ASHRAE, the Illuminating Engineering Society of North America (IES), IESNA, the International Association of Lighting Designers (IALD), the International Code Council, ANSI, the U.S. DoE, the National Institute of Standards and Technology (NIST), NEMA, and the Cybersecurity Council (standards, guidelines, and best practices related to grid cybersecurity). There is also a number of industry groups and initiatives. To list one example, the International WELL Building Institute™ (IWBI™) is an advocacy entity that produces quality-of-work-environment standards for interior spaces, buildings, and communities that strive to implement, validate and measure features that support and advance human health and wellness. The WELL Building Standard™ (and the follow-on WELL v2™) is a rating system focused on the ways that buildings and their contents, can improve comfort, drive better choices, and enhance occupants' health and wellness [51]. Wellness issues addressed in the standard include air quality, water quality, thermal comfort, lighting, sound and noise, space adequacy for the work function at hand, and optimal movement facilitation. In the context of lighting according to IWBI, the "WELL Light concept" aims at creating lighting environments that are optimal for biological, mental and visual health.

Lighting Power Density (LPD) is defined by ANSI, ASHREA and IESNA as the load of any lighting equipment in a specified area; it is described as watts per square foot (W/sq. ft. or W/ft^2) of the lighting equipment, for a given occupancy/space type (e.g. office, hospital, museum and granularly for work spaces, restrooms, staircases, and so on). Effectively it is the maximum allowable lighting density—specifically, the lighting power allowance (LPA)—permitted by a given city, state, or national code. A basic calculation method is the "Space by Space Method" where the specification enumerates a list of various space types within a building and the associated watts per square foot allowance. For example, the 2018 IECC code specifies an LPD 0.79 W/ft^2 for office space and more specifically (for example) an LPD of 0.62 W/ft^2 for a lounge/breakroom rooms, 0.85 W/ft^2 for a restroom, and 0.81 W/ft^2 for an open plan office.

Two (of many) applicable standards are the International Energy Conservation Code (IECC) developed by the International Code Council and ASHRAE (ASHRAE 90.1).

- IECC is updated tri-annually, the latest being the 2018 version (original version was issued in 2009); the 2015 version is currently adopted in many states in the U.S., particularly states with large urban environments. For Commercial Lighting the code includes, among other aspects, luminaire level lighting control (LLLC) and interior lighting power limits (also detailing exceptions) (external lighting guidance is also included). Section C405.2 mandates that lighting must use system control or LLLC; controls must monitor occupancy and ambient light and must include multiple set points and zoning. Exceptions are provided for security, emergency, and exit areas. Basic interior system capabilities (for example, as identified by the DesignLights Consortium® which aims at fostering broad adoption of high-performing commercial lighting solutions [52]) include the following functionality: networking of luminaires and devices; luminaire and device addressability; zoning; occupancy sensing; daylight harvesting/photocell control; high end trim, and continuous dimming; BMS integration is also desirable, when achievable. The code mandates separate lighting controls for open offices, Auto-OFF for occupancy sensors reduced to 20 min, and additional areas where occupancy sensor controls are required (compared with the 2015 version of the code—e.g., in enclosed offices, open plan office areas, and meeting rooms). As a minimum, the digital control system is required to include the following: luminaires must be capable of continuous dimming; luminaires must be capable of being addressed individually or if as a group not more than 4 in the group; not more than 8 luminaires can be controlled together in a daylight zone. Fixtures must be able to be controlled through a digital control system that includes the following functionality: control reconfiguration (including for occupancy sensors) must be based on digital addressability; load shedding mechanisms must be provided; individual user control of overhead illumination in open offices must be provided [53]. Regarding LPD, C 405.3.2(2) provides a list of building area types and common space types. Examples of building area types include dining cafeteria (LPD 0.79 w/ft^2); exercise center (LPD 0.65 w/ft^2); health care clinic (LPD 0.82 w/ft^2); hospital (LPD 1.05 w/ft^2); museum (LPD 1.06 w/ft^2); private office (LPD 0.79 w/ft^2); parking garage (LDP 0.15 w/ft^2). Examples of LPD for common space types were listed just above. For example, for an open plan office of 20,000 sq. ft., the compliance wattage would be 20,000 × 0.81 = 16,200 W.

- ASHRAE 90.1. This standard, updated every 3 years, last updated in 2016, defines the minimum set of requirements for energy efficient design of buildings. Among other matters, it defines interior lighting power limits for commercial lighting, exterior lighting power limits for commercial lighting, and dwelling unit efficacy [53]. The LPDs generally align with the values defined in IECC 2018. It also mandates occupancy sensing, energy monitoring, primary and secondary daylight zones, and (reduced) timeout settings. Per the standard, occupancy sensors must be installed in conference rooms, break rooms, private offices, restrooms, and storage rooms. Measurement systems must be installed in new buildings to individually monitor energy use of HVACs, interior lighting, exterior lighting, and receptacle circuits. The usage data must be captured every 0.25 hours and must be posted hourly, daily, and monthly. Space control requirements include local control (manual ON/OFF), automatic daylight controls, automatic partial-OFF when space is 50% unoccupied (open plan office space lighting allowed to turn on automatically to more than 50% with partial occupancy).

Other relevant standards include the following:

- Standards and guidelines for office and industrial/factory illumination, including standards from IES (e.g., The Lighting Handbook, tenth Edition), OSHA (e.g., Subdivision J, 437-002-0144, Additional Oregon Rules for General Environmental Controls), ANSI (e.g., ANSI/IESNA RP-1-12 American National Standard Practice for Office Lighting, and ANSI/IESNA RP-7-01 Recommended Practice for Lighting Industrial Facilities), Canadian Centre for Occupational Health and Safety, and other entities worldwide.
- NEMA/ANSI C137.3-2017, Minimum Requirements for Installation of Energy Efficient Power over Ethernet (PoE) Lighting Systems, documents installation requirements. In particular, the standard recommends the minimum wire diameter (gauge) to limit resistive line losses to less than 5% of the total power delivered, postulating a 50 m average cable length.
- UL has established the UL Cybersecurity Assurance Program (UL CAP) to address the cybersecurity concerns in CLSs. UL CAP evaluates the cybersecurity of network-connectable products and systems; additionally, it evaluates the processes utilized to develop and maintain products and systems in a secure manner, for example, how software updates are promulgated and managed. UL CAP uses UL 2900-1 cybersecurity standard as the baseline. UL CAP services and software security efforts are consistent with the U.S. White House Cybersecurity National Action Plan (CNAP).
- There are several recommendations for evaluating cybersecurity risks, including the NIST Cybersecurity Framework 1.1, the NIST 800 series, the International Electrotechnical Commission 62,443 series, the ISO 27000 series, and UL 2900-1 just cited. However, there is no mandatory requirement for cybersecurity certification for connected lighting systems at this juncture.
- Additionally, an Open Connectivity Foundation (OCF) specification known as OCF 2.0 was under development at press time for eventual submission to ISO/IEC JTC 1 (Joint Technical Committee for ICT standardization of the ISO and the International Electrotechnical Commission [IEC]). OCF 2.0 incorporates a security framework based on Public Key Infrastructure (PKI) and also incorporates cloud management capabilities. In this context, security protocols deal with access control lists, which define the levels of access that are granted to a client with certain credentials. Procedural trust allows users to endow devices certain credentials through a process called onboarding and provisioning. Ultimately one needs to establish if one can trust a given device to connect to the network.

Smart Lighting is also seeing deployment in the residential smart home context. Using cloud services to connect with smart home products such as Amazon Alexa, Google Home, Nest, and Ring, consumers can schedule and trigger lights under a variety of conditions, for example when the Ring doorbell is pressed. One of the many possible IoT approaches for lighting, or more generally for generic device control, is the use of the set of (Universal Plug and Play) protocols. UPnP-based and other home area network-based OCF specifications can standardize the control of devices such as smart switches, dimmers and outlets. The OCF 1.0 specification has been ratified as an International Standard by ISO/IEC JTC 1 as ISO/IEC 30118 (Parts 1-6). These OCF resource descriptions enable the industry-specific deployment of IoT-enabled devices (including lighting) for Smart Home, Healthcare, Automotive and Commercial Buildings.

Standardized device and service descriptions for the Lighting Controls have been published by OCF (and the predecessor entity, the UPnP Forum) as V1.0 Standardized Device Control Protocol (DCP) as early as 2003. The DCPs include:

- BinaryLight V1.0 (2003);
- DimmableLight V1.0 (2003);
- Dimming V1.0 (2003);
- SwitchPower V1.0 (2011).

For example, the BinaryLight:1 Device Template Version 1.01 for UPnP™ Version 1.0 (2003) describes a device template that is compliant with the UPnP Architecture Version 1.0 and Version 1.0 of the UPnP Standard Device Template. The specification defines a device type referred as to BinaryLight:1 which is a simple representation of a bulb or other kind of light emitting device that can be switched completely on or off. The BinaryLight:1 specification provides the following (simple) functionality: *Switching the light source on or off.*

UPnP mandates that devices, such as a lighting entity, support IP-based addresses, and basic protocols including Hyper Text Transfer Protocol (HTTP) (with intrinsic Extensible Markup Language [XML] syntax), User Datagram Protocol (UDP) (on port 1900), and Transmission Control Protocol (TCP). Refer back to Fig. 3.5 for the requirements of the protocol stack. Although UPnP has been utilized more extensively in the smart home context than the commercial smart building or smart factory context, it is still a usable technology for the latter cases. UPnP utilizes these protocols to advertise the devices' presence and for information transfers. Some elements act as end-point devices while others act as Control Points – a Control Point is typically a multifunction/data-processing IoT gateway, which resides on the Local Area Network and interacts with the rest of the devices in the local environment via a designated router, on a specified IP subnet. The basic functions of the UPnP arrangement are as follows: device addressing, service discovery, device description, device control, and event notification ("eventing"). Figure 3.17 depicts graphically these processes.

As noted, for addressing UPnP utilizes the IP addressing scheme (particularly IPv4). When a device first joins the network (Step 0) it acts as a Dynamic Host Configuration Protocol (DHCP) client to assign itself an IP address; then it searches for a DHCP server. If no DHCP server is located, the device assigns itself a unique IP address from a set of reserved private addresses, specifically 169.254.0.0/16 (similar to the method described in RFC 3927). The next process is (service) discovery (Step 1). When the device is added to the network, UPnP allows the device to advertise its higher layer services to other devices on the network using the Simple Service Discovery Protocol (SSDP); this is done by broadcasting *SSDP alive* messages utilizing HTTPMU (HTTP multicast over UDP). SSDP also enables a device to

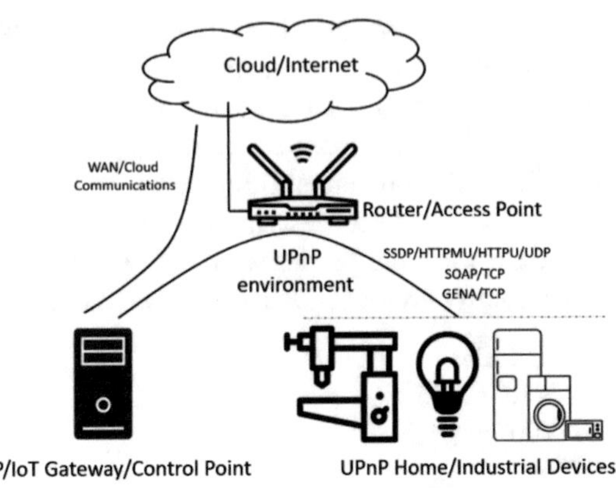

1a: Discovery: sends response to device advertisements with SSDP / HTTPU / UDP unicast
2: Solicits service descriptions with SOAP / TCP using XML
3: Issues controls to devices using SOAP / HTTP / TCP
4: If subscribed, gets events changes compared with state tables using GENA / XML

Devices run HTTP server

0: DHCP Client or AutoIP
1: Discovery: device multicasts: SSDP / HTTPMU / UDP service advertisements
1a: Discovery response: unicast from control point SSDP / HTTPU / UDP
2: Description: XML / SOAP messaging to Control Point Architecture Schema; Working Committees definitions of device types and services; Vendor information (e.g., name, model)
3: Control from Control Point: SOAP / HTTP / TCP
4: Eventing: GENA framework/ XML / TCP against State Table

Fig. 3.17 The UPnP process for device control

passively listen to *SSDP alive* messages from other devices on the network. When two devices discover each other, typically an end device and a Control Point, a discovery message is exchanged in a unicast manner (with SSDP over HTTPU [HTTP Unicast]). This message contains basic device information such as the device type and its services and capabilities.

An important step (Step 2) is a determination of the device description. After the devices discover each other they exchange support information such as manufacturer identification, device model name or number, website URL of the manufacturer where additional device service information is available, parameters or data to be exchanged between the device and the Control Point for a given service, and so on. The information is exchanged in XML format using the as Simple Object Access Protocol (SOAP). SOAP is a standardized method to run remote procedure calls (RPCs); the "object" in question is some appropriate software module. SOAP runs over HTTP and uses XML to describe remote procedure calls to a server and return results from those procedure calls.

The third step (Step 3) in the UPnP-based environment is "control": a device or program (such as the Control Point) can instruct another device or program (e.g., the terminal appliance, such as a lighting device) to perform a specified action; this is done using SOAP over TCP. The control can be direct, such as the Control Point directing the device to do something; or the control can be more sophisticated where, for example, after obtaining information about the device and its services, the Control Point endeavors to utilize the service described by the URL provided by the manufacturer and acquired during the device description step (also using SOAP—as noted similar to a programming function call) to secure additional capabilities to effectuate the control of the device. Requesting a service is done by sending a SOAP request to the "control URL" of the Control Point, with the appropriate set of parameters. For many programming languages, libraries are available that can be utilized to implement SOAP requests and process SOAP responses.

In the context of Step 4, eventing, General Event Notification Architecture (GENA) is the mechanism used for event notification in UPnP; TCP is the underlying transport mechanism and messages are coded in the XML format. UPnP devices maintain "state variables" which are utilized for keeping state information in the devices. A Control Point or other entities (programs) can subscribe to state changes for a device, a process known as eventing: when a state variable is changed, the new state is sent to all devices and entities (programs) that have subscribed to the event.

3.4.8 *Connected Lighting Systems Evaluations*

Since 2017 (and through 2021) the DOE Next Generation Lighting Systems (NGLS) program, managed by DOE's Office of Energy Efficiency and Renewable Energy, has been evaluating connected lighting systems for both indoor and outdoor applications. The connected lighting systems are being evaluated by the NGLS program for such features as installation ease, configuration ease, control operation, lighting quality, energy savings, and user satisfaction. The results were planned to be made publicly available to benefit lighting technology developers, specifiers, installers, and users [10]. NGLS outdoor evaluations started in 2019; several outdoor parking lot systems were installed to be evaluated in real-life settings with the goal of identifying possible product improvements.

3.4.9 *IoT Support of Lighting*

In a smart building, support systems (lighting, HVAC) are connected and can interact; in IoT-based smart buildings a large set of sensors are affiliated with the support systems and the collected data can be analyzed in (near) real time to affect a variety of management and participants' decisions [54]. One of the many applications of IoT in smart buildings relates to space utilization (others might include environmental control, indoor positioning, occupant tracking, social distancing, and asset tracking – these last three especially in the context of beacons, tags, and/or smartphones); lighting is a corollary to space utilization in terms of occupancy/vacancy.

3.4.10 Lighting Arrangements

As noted, LED technology has reduced the power required for lighting applications, while advances in the PoE field have increased the amount of power that can be delivered to a networked device over a single cable. Several major LED luminaire manufacturers have introduced PoE-based lighting systems in recent years, making this combination a potentially disruptive technology. PoE-based lighting systems utilize drivers that utilize the PoE-defined power supply, therefore the implementer cannot just use any fixture/luminaire in a PoE installation; the LED drivers must be able to be powered from Ethernet cables connected to a network switch—these drivers are designed for low-voltage DC supply while the generic LED driver is designed for line-voltage AC supply. Additionally, given that most modern commercial fixtures typically incorporate sensors, the interfaces to these sensors both for power and signal transmission must conform to the PoE standard. As of press time PoE light product line is relatively limited. However, manufacturers are developing an increasing number of PoE product suites and are cautiously viewing the incremental acceptance of this technology in the marketplace, while retaining significant investment in traditional LV technologies.

A high level PoE-based CLS architecture entails a Device Layer (luminaires and sensors), a Communication and Power Layer, and a Management Layer (see Fig. 3.18).

- Device Layer
  - Devices (Luminaires)
  - Sensors and Local Controls (wired or wireless) (e.g., occupancy sensors, ambient light sensor)
  - Infrastructure Elements (cabling, PoE gateways, drivers, control elements) (gateways: unregulated voltage output [connected lighting], regulated voltage output [constant voltage LED lights, linear LEDs], wireless gateway [zone relays])
- Communication and Power Layer
  - Network Switches for Data and Power (PoE switches)
- Management Layer
  - Application Software (control, manage, reporting)

In a practical environment, there will likely be a hybrid design that includes LV and high voltage devices with wired or wireless control technology that uses IP or traditional communication.

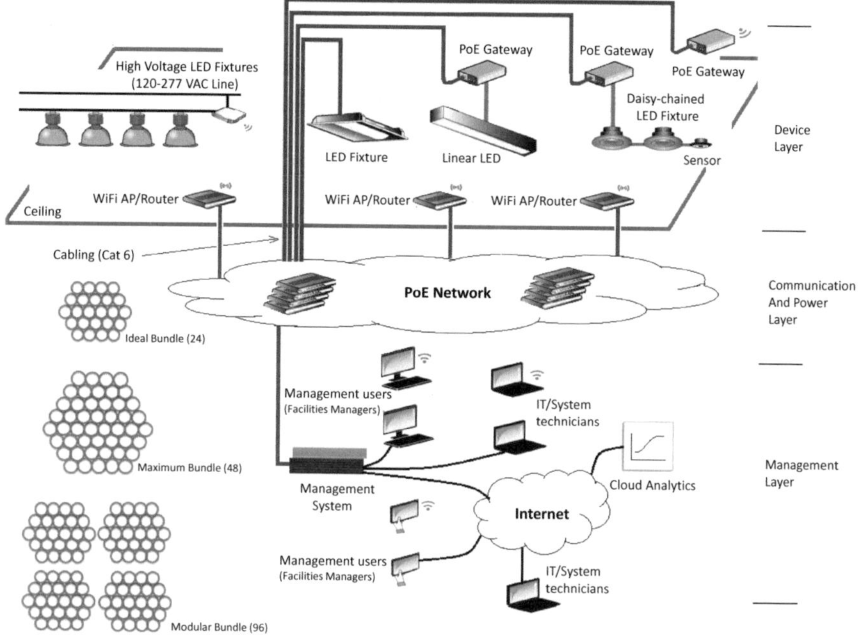

Fig. 3.18 High level PoE-based lighting architecture

3.4.11 *Outdoors/Street Lighting Considerations*

There are in the range of 30 million street lights in the U.S. alone. In addition, there are many parking lots, including outdoor parking lots for commercial buildings and industrial buildings. The energy savings that result from a conversion to LED light are substantial (reducing the consumption to approximately one-third of the current electricity consumption). Some considerations about street lighting include the following: safety; energy efficiency; pedestrian-friendliness; visual comfort (appropriate CCT and quality fixture design); minimal glare, light trespass and skyglow (e.g., high CCT increases the perception of glare/light trespass/skyglow); minimize ecological and circadian-disruption impact [55]. Using appropriate Back-lit, Upright, and Glare (BUG) rated fixtures reduces adverse impact and improves visual comfort. The BUG rating was developed by the IES and The International Dark Sky Association in order to quantify the light escaping in undesired directions from an outdoor light fixture. *Backlight* is the light that is dispersed from behind the fixture into areas where it is unwanted (specifically opposite area to the area where light is intended to be); it impacts light trespass on adjacent sites; street lighting can cause light trespass if not designed properly and is the most common complaint by the public. *Uplight* is the light dispersed above the top of the fixture; it creates light pollution and sky glow. *Glare* is the amount of front light in the forward zones when light is directed into eyes, not the general target area—this occurs when the light is too strong or concentrated creating a safety issue and dramatically degrading visibility. Glare is perceived as discomfort or disability, especially affecting older people; it necessitates increased lighting levels on the target zone to compensate its negative effects.

Different luminaires have different BUG ratings. BUG is described in IES TM-15-11. The defined BUG zones are as follows (see Figs. 3.19 and 3.20):

- Backlight Sub-Zones
  - BVH: Backlight Very High (covers 80–90° where degrees are measured in semi-circle from the ground to the maximum elevation above fixture)
  - BH: Backlight High (covers 60–80°)
  - BM: Backlight Mid (covers 30–60°)
  - BL: Backlight Low (covers 0–30°)
- Uplight Sub-Zones
  - UH: Uplight High (covers 100–180°)
  - UL: Uplight Low (covers 90–100°)

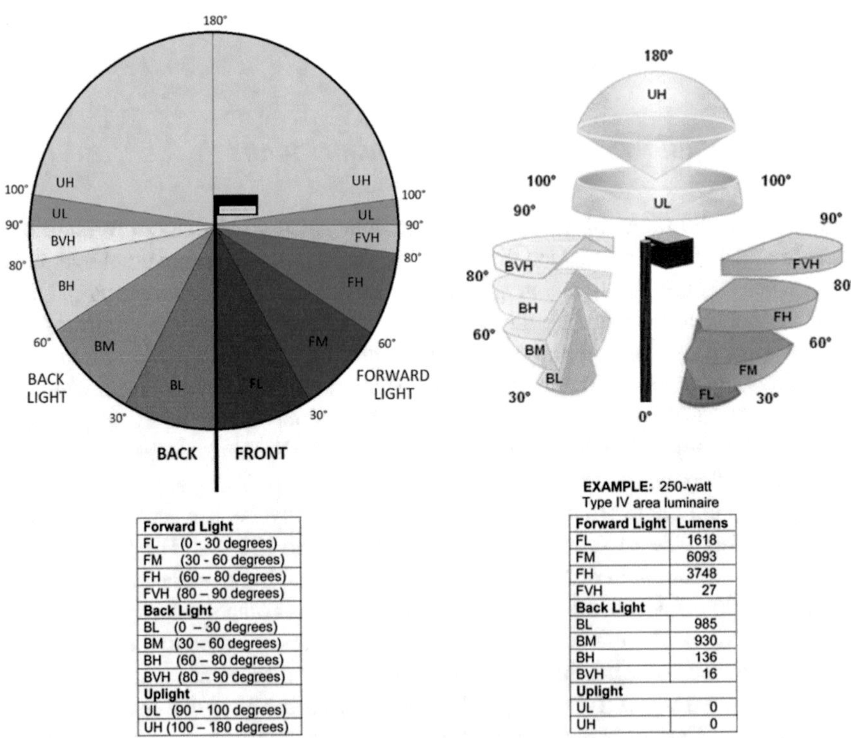

Forward Light
FL (0 - 30 degrees)
FM (30 - 60 degrees)
FH (60 – 80 degrees)
FVH (80 – 90 degrees)
Back Light
BL (0 – 30 degrees)
BM (30 – 60 degrees)
BH (60 – 80 degrees)
BVH (80 – 90 degrees)
Uplight
UL (90 – 100 degrees)
UH (100 – 180 degrees)

Forward Light	Lumens
FL	1618
FM	6093
FH	3748
FVH	27
Back Light	
BL	985
BM	930
BH	136
BVH	16
Uplight	
UL	0
UH	0

Fig. 3.19 BUG model—IES TM-15-11

Subzone		B0	B1	B2	B3	B4	B5	Example (*)
Backlight/Trespass	BH	110 lm	500 lm	1000 lm	2500 lm	5000 lm	>5000 lm	500 lm
	BM	220 lm	1000 lm	2500 lm	5000 lm	8500 lm	>8500 lm	1000 lm
	BL	110 lm	500 lm	1000 lm	2500 lm	5000 lm	>5000 lm	1000 lm

(maximum zonal lumens)

Secondary Solid Angle		U0	U1	U2	U3	U4	U5	Example (*)
Uplight/Skyglow	UH	0 lm	10 lm	50 lm	500 lm	1000 lm	>1000 lm	0
	UL	0 lm	10 lm	50 lm	500 lm	1000 lm	>1000 lm	0

(maximum zonal lumens)

Secondary Solid Angle		G0	G1	G2	G3	G4	G5	Example (*)
Glare/Offensive Light	FVH	10 lm	100 lm	225 lm	500 lm	750 lm	>750 lm	100 lm
	BVH	10 lm	100 lm	225 lm	500 lm	750 lm	>750 lm	5000 lm
	FH	660 lm	1800 lm	5000 lm	7500 lm	12000 lm	>12000 lm	100 lm
	BH	110 lm	500 lm	1000 lm	2500 lm	5000 lm	>5000 lm	110 lm

Glare Rating for Asymmetrical Luminaire Types (Type I, Type II, Type III, Type IV) (maximum zonal lumens)

Secondary Solid Angle		G0	G1	G2	G3	G4	G5
Glare/Offensive Light	FVH	10 lm	100 lm	225 lm	500 lm	750 lm	>750 lm
	BVH	10 lm	100 lm	225 lm	500 lm	750 lm	>750 lm
	FH	660 lm	1800 lm	5000 lm	7500 lm	12000 lm	>12000 lm
	BH	660 lm	1800 lm	5000 lm	7500 lm	12000 lm	>12000 lm

Glare Rating for Quadrilateral Symmetrical Luminaire Types (Type V, Type VS) (maximum zonal lumens)

(*) Example for a BUG rating B2 U0 G2 luminaire *Types (Type I, Type II, Type III, Type IV)*

Fig. 3.20 IES TM-15-11 BUG ratings

- Glare (Front Light) Sub-Zones
  - FVH: Forward light Very High (covers 80–90°)
  - FH: Forward light High (covers 60–80°)
  - FM: Forward light Mid (covers 30–60°)
  - FL: Forward light Low (covers 0–30°)

BUG is used in conjunction with the International Dark Sky Association's light zones (LZs), which are accepted levels of light in outdoor areas:

- LZ0: No Ambient Lighting
- LZ1: Low Ambient Lighting
- LZ2: Moderate Ambient Lighting
- LZ3: Moderately High Ambient Lighting
- LZ4: High Ambient Lighting

The *ANSI/IES RP-8-18: Recommended Practice for Design and Maintenance of Roadway and Parking Facility Lighting* is a major 2018 revision to previous versions of ANSI/IES RP-8, being an aggregation of several IES Standards covering roadway and parking facility lighting. ANSI/IES RP-8-18 is a compilation of lighting design techniques and criteria and is intended to be a single source of reference for roadway lighting. This Practice enables designers to [56]:

- Improve visual quality for motorists
- Deploy quality light and increased contrast to identify hazards
- Minimize environmental impacts and overspill of night lights
- Utilize lighting systems that are easy to maintain
- Utilize lighting systems that minimize energy use

The use of a properly-rated BUG fixture is needed to reduce glare, uplight, and light trespass. Additionally, installing controls is highly desired, given that the ability to dim as described in RP-8-18 can save 50% in energy costs. Sensors of the environment are required in order to support dynamic lighting control; importantly the intrinsic ability to achieve controls requires (IoT) communication (wired and/or wireless) to a centralized management system. Remote light poles can also support additional functionality such as acting as a 5G microcells, acting as a Wi-Fi Access Points, and providing last-hop (wireless) access to a backhaul Fiber-To-The-Node system. Other applications include video surveillance and air quality monitoring.

3.5 CLS Market Considerations

There are various CLS market forecasts spanning a wide range. The market for smart lighting technology is inherently much larger than the PoE lighting market. The overall CLS market (including LV and PoE) has been estimated at approximately USD 25 billion by the end of 2023 by some [57], USD 27 billion by 2024 with a CAGR of over 20% during the forecast period by others [58], and USD 30.85 billion by 2025 by yet others [59] (the global Smart Lighting market was valued at USD 6.56 billion in 2017 in [59], USD 12.1 billion in 2016 in [60]). Europe has been the largest smart lighting market, accounting for more than 35% of global revenue while sales in Asia-Pacific are expected to witness the fastest growth [58]. The commercial segment of the PoE Lighting market is projected to be valued at USD 1.27 billion by 2026 according to some researchers [61], or USD 8.24 billion by 2025 according to other researchers [62]. Clearly there is a wide disparity in these numbers and the true value is likely somewhere in between; the market for PoE lightning is dominated by new construction only, not generic space retrofit.

The smart lighting market can be taxonomized into lighting sources, communication medium, products, applications, and services. Smart Lighting encompasses sources such as LEDs, Fluorescent Lamp, and High Intensity Discharge Lamp (HID). As noted previously, control connectivity can be wired (e.g., DALI, PLC, PoE) or wireless (e.g., Wi-Fi, Bluetooth, Zigbee). At press time the smart lighting

Table 3.4 Forecast PoE number of ports

Year	Global ($M)	U.S. market ($M)	Number of fixture/ports at $600 inclusion in market report analysis	Aggregate office space supported (at 2000 lumens per luminaire and 25 lumens per square foot)
2018	450.00	153.00	255,000	20,400,000
2019	506.70	172.28	287,130	22,970,400
2020	570.54	193.99	323,308	25,864,670
2021	642.43	218.43	364,045	29,123,619
2022	723.38	245.95	409,915	32,793,195
2023	814.53	276.94	461,564	36,925,137
2024	917.16	311.83	519,721	41,577,705
2025	1032.72	351.12	585,206	46,816,496
2026	1162.84	395.37	658,942	52,715,374

ecosystem includes, but is not limited to, vendors such as Acuity Brands Inc., Bridgelux (U.S.), Control4 Corporation, Cree, Daintree Networks, Digital Lumens Inc., Echelon Corporation, General Electric, Honeywell, Hubbell Lighting Inc., Ingersoll-Rand, Koninklijke Philips N.V., Legrand, Lutron Electronics Co. Inc., Osram Licht AG, Schneider Electric, Siemens, SoftDel, Streetlight Vision, Zumtobel Group AG. Early players in the PoE lighting market include, but are not limited to, Signify N.V., Cisco Systems, Inc., IGOR Inc., Molex Incorporated, Leviton Manufacturing Company, Inc., NuLEDs Inc., Platformatics, Cree, Inc., Axis Communications AB.

The following quote represented the press time environment for PoE-based lighting (also see Table 3.4):

> As the market is still at inception, companies are wary of using it for lighting. Even though PoE is successful in the telephone system, it is still not accepted widely in lighting. Few companies realize its benefits and partnerships have been formed between lighting and technology companies. Plans are to integrate these PoE lightings in the buildings, especially in commercial sectors. However, the limitations on the power delivered by PoE-based power sourcing to end devices are hindering the growth of the market [61].

3.6 Design and Implementation Considerations

As implied earlier, smart lighting is fully possible using line voltage to power the luminaires and wireless for the controls; however, this section focuses on PoE-based lighting systems. As we noted in Sect. 3.4, in the recent past one has experienced a convergence between LED power requirements (decreasing power with improved luminous efficacy) and PoE power capabilities (increasing power with newer IEEE standards): the first PoE standard in 2003 (IEEE 802.3af-2003) supported up to about 15 W DC to a load; the revised 2009 standard called PoE+ (IEEE 802.3at-

2009) is rated to 50 W, and PoE++; the newest standard called IEEE 802.3bt-2018) can deliver (in principle) nearly twice the power achievable with PoE+. Note, in passing, that existing, in-place Ethernet wiring is not sufficient for PoE++; instead, a new backward-compatible wiring and connector scheme must be utilized [63]. The availability of PoE with higher power ratings can eliminate the need for separate DC wiring to the luminaire or fixture; however, if the system is not properly designed, the efficiency is impacted and offset by the increased losses associated with increased voltage drop in the Ethernet cables and connectors. Additionally, ceiling/conduit temperature will vary, impacting the performance of various system components including that of the PoE switch, LED driver, and the conductor DC resistance (DCR) [64].

In the context of PoE-based systems, a practical consideration (in key metropolitan areas in the U.S., but with comparable concerns in a number of other locations) is who actually would install a PoE-based lighting system: traditional electricians that typically install line-voltage equipment (e.g., International Brotherhood of Electrical Workers—IBEW), or electricians/technicians that typically install LV LAN/data/signal equipment (e.g., Communications Workers of America—CWA). The PoE voltage is below the typical electrical-code threshold limit of 48 V, therefore a licensed electrician is not needed to install the wiring, thus possibly reducing installation or re-wiring costs, installation time, and cabling costs—LV electricians/technicians occasionally charge less than line-voltage electricians and that was an initial motivation for PoE-based systems. However, up to the present, architectural light fixtures have invariably been installed by electricians authorized and qualified to install line-voltage systems [50]. In pragmatic terms, the original driver to possibly avoid using union labor (e.g., IBEW) and enjoy lower installation cost with other entities (e.g., CWA or non-union shops) is being lost in many jurisdictions. For example, in New York City the press-time rules were that IBEW must hang the luminaires; furthermore, emergency lighting must remain line-voltage; no changes to the city building codes were on the immediate horizon. Therefore, the pertinent question is what happens when PoE fixtures and network switches are delivered to the site in question? Likely, line-voltage electrician will mechanically install the fixture and physically connect the Ethernet cable to the fixture; the remaining upstream work (connecting the cable to the switch in the plenum or in the IT closet) will be undertaken by a system integrator or LAN contractor.

3.6.1 Technical Considerations for Building Applications

Energy losses in Ethernet cables utilized between luminaires and PoE switches in PoE-connected lighting systems are a major technical concern: while PoE lighting systems offer in principle better energy efficiency compared to traditional line-voltage-based AC systems, there is a tradeoff dictated by increased voltage drop in the LV Ethernet cabling. Cable features that affect its energy performance include:

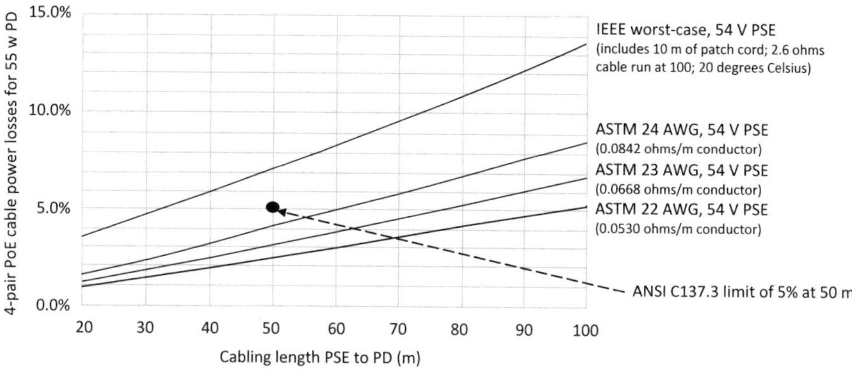

Fig. 3.21 Impact of cable selection on power losses

Category (e.g., 5e, 6 , 6a), gauge, fire rating, and shielding approaches. Cable losses decrease when increasing the conductor diameter, i.e., with numerically smaller American Wire Gauge (AWG) rating. See Fig. 3.21. Besides using conductors of numerically smaller AWG designation (i.e., larger diameter), cable power and energy losses in twisted-pair cables can be reduced by reducing the length of cabling from PSE to PD (i.e., the link section), reducing the number of conductor twists per unit cable length (thereby reducing conductor length), or using higher-conductance (purer copper) material [64]. Resistive line losses in a PoE lighting distribution system utilizing CAT5/6 cabling can be more than 15%; it clearly follows that when resistive line losses are high the system-level efficiency will be significantly lower than the efficiency of traditional AC-powered systems. Cable losses are inversely proportional to the conductor's diameter; thus, utilizing cabling with an appropriate wire gauge will limit resistive losses and enable a PoE lighting system to match or exceed the energy efficiency of a traditional AC lighting system in "real life" settings in commercial buildings and factories.

To guide the design of PoE lighting systems, ANSI recently published ANSI C137.3-2017, which specifies the minimum AWG for UTP Ethernet cables as a function of the power required and/or dissipated by the PD. Specifically, ANSI C137.3-2017 aims at limiting cable losses to less than 5% of PSE output power over a postulated average cable length of 50 m for cables varying in AWG, category, shielding, and manufacturer. The specification recommends limiting PD input power to 55 W for 4-pair PoE over 24 AWG cabling, including the conductor cable, the patch cords and the connectors—power losses are greater when cables are connected to patch cords, bundled in conduit, and loaded with PDs approaching 90 W input power. Stated differently, ANSI C137.3-2017 specifies the minimum AWG rating for installation of energy efficient PoE Lighting Systems using UTP Ethernet cables as a function of power utilized by the PD; it bases its guidance on an included table of nominal DCR values, which are in turn derived from data published in American Society for Testing and Materials (ASTM) B258-14 (ASTM 2014) for straight solid conductors at 68 °F.

Although cable power losses do not increase by cable bending or bundling in uninsulated conduit, environments with higher ambient temperatures experience greater power losses due to increased conductor DCR. In addition, product selection and installation practices will have assessed criticality when PDs based on the IEEE Standard 802.3bt will operate at 90 W input power conveyed by a single Ethernet cable. While the standards published by IEEE and by the TIA in this arena limit DCR mostly for performance, safety guidelines are established by the National Fire Protection Association (NFPA) as NFPA 70 [64, 65].

The issue has been assessed by the U.S. DoE (among others) [64, 66]. Extensive testing was undertaken in DOE's Connected Lighting Test Bed in 2017 and 2018, aimed at establishing the impact of cable choices and installation practices, also in conjunction with current guidelines. The DoE analysis established that with 44 W luminaires and room temperatures less than 86 °F, cable power losses did not increase by the practice of bundling the cables in uninsulated conduit or bending the cables; however, environments with higher temperatures exhibit greater power losses. Consequently, there is a need for PoE lighting system vendors to state what the minimum AWG must be to meet ANSI C137.3's guidance.

The deployment of smart lighting in new construction situations is relatively simple from a planning perspective: the designer can specify the use of the appropriate cabling and related systems that meet the recommended standards from the onset. Retrofit situations are more complex. In these situations, one can use form/fit/function LED replacements for traditional lighting in existing sockets, luminaires and troffers (the rectangular light fixture that plugs into a modular dropped-ceiling grid) with the necessary AC/DC converter built into the LED-based bulb itself or a separate converter mounted in a housing; connectivity for control is achieved via some other mechanism [63]. Otherwise, the AC/DC converter has to be located in hidden fashion within the luminaire or fixture or in a nearby closet. In the latter case, new wiring is needed so that power can supplied via dedicated DC power rails – PoE approaches support this arrangement by providing DC power, but up to the maximum current rating (per Table 3.3). Here control is achieved via the Ethernet cable.

3.6.2 Design Considerations for Building Applications

Installing PoE in general and PoE-based smart lighting in retrofit applications is not simply a matter of connecting a DC-power supply to the in-situ Ethernet cabling: special cabling and coupling techniques are needed. Practitioners recognize that using PoE for SSL and/or smart lighting will not be a quick, drop-in replacement [63]. Although PoE's low voltages implies that a licensed electrician is not necessarily needed, the higher power rating and current level of 802.3bt Type 4 give rise to new considerations related to self-heating due to resistive losses in the cable. Cable heating engenders the possibly of self-ignition of ceiling materials in contact with the cables, especially when cables are bundled tightly and run as

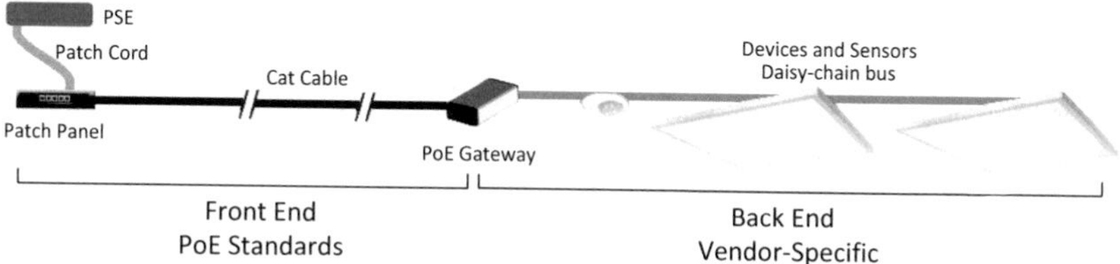

Fig. 3.22 Typical hybrid mix of PoE and vendor-specific cabling

aggregates in cable troughs, all of which reduces airflow and increases self-heating. The possibility of in-wall smoldering has driven the development of strict codes for AC-line wiring since every junction is a potential high-resistance point which can overheat and ignite nearby materials, thus dictating that all AC connections must be in enclosed, fire-resistant boxes. In addition, many PoE lighting system also need to rely on vendor-specific cabling in the "back end", as depicted in Fig. 3.22.

3.7 Examples of Products and Element Costs

One of the questions at this juncture is if smart lighting will be deployed commercially using wireless control or wired control via PoE. Probably one will see a mix of the two, but the interesting follow-on question is what is the percentage split? Clearly the commercial success of a technology is driven in large measure by the financial advantages of the new paradigm—Return on Investment—more so than the "wonders" of the new technology. For example, Voice over IP has been a great success because of the obvious reductions in cost of service delivery and the reduced in-premises expenditures when replacing vendor-specific high complexity PBXs with low-footprint off-the-shelf server hardware running soft switch software. The same reductions in cost when using LED lighting are clearly demonstrable, with relatively short breakeven intervals to offset the initial investment. The financial advantages of connected lighting, and more specifically of PoE-based lighting remain to be fully documented at press time. Clearly, there are different considerations for a greenfield environment, especially with high-end urban and/or Class A office space (e.g., Hudson Yards office development in New York City)—especially aiming at occupancy by high-tech firms—compared with retrofit environments, suburban space and/or existing Class B and C office space, or factory floors.

It is also important that for both LED illumination as well as for PoE-supported illumination, a full product line of bulbs and luminaires be available, being that real estate owners, architects, and lighting designers, as well as tenants are not looking for a "proof of concept" where a handful of bulbs are available, but a full suite of lighting fixtures, bulbs, and arrangements are available. For example, there typically are workspaces and common spaces; these require different illumination from both a

Table 3.5 Practical cost considerations

Element	Technology LED with LV	LED with PoE	Fluorescent
Fixture	$350	$300	$200
Dimming ballast and lamp			$75
Fixture installation	$250	$250	$250
PoE cable and labor		$150	
PoE switch port		$75	
Sensors/control	$200	$75	$200
Total	$800	$850	$725

functional perspective and from a light emission perspective. Considerations include but are not limited to daylight requirements in regularly-occupied spaces, daylight requirements in common spaces, tunable lighting solutions, lighting intensity modulation, solar glare management, glare from other lights (aiming at a Unified Glare Rating [UGR] of 19 or lower), shielding, color rendering quality, etc. These considerations drive the choice of luminaires, fixtures, lightbulbs, and placement. Obviously, all of these factors impact employee productivity in both an office environment as well in a factory (or warehouse) setting.

PoE costs were not overwhelmingly compelling (as in the case of LED vs. traditional fixtures) at press time and the availability of a panoply of products was still lacking. In addition, the life-cycle cost implications – including day-to-day facility's operation, and maintenance – of PoE-based lighting are not yet well understood at press time.

The product line needed to support the working space of a modern office building might need the following types of lighting fixtures: pendant mount LED linear, recessed LED downlight, surface mount LED cylinder, recessed LED downlight, surface mount LED linear, pendant mount LED ring, recessed LED linear, perimeter mount LED linear, pendant mount LED decorative, surface mount LED strip, surface mount track, LED track head, wall mount LED decorative, city approved edge lit exit. This implies that to be practical, PoE lighting needs a wide array of fixtures and luminaires. Average retrofit costs at press time are shown in Table 3.5 for a representative retrofit application. Although PoE-based lighting was not financially a "killer-app", it does make control easy to achieve. These costs will undoubtedly change in the future.

3.8 Implementation Considerations

Recent studies have shown that the savings of installing a PoE-based connected light system are relatively small. Implementation savings of 4–8% have been documented by some for large offices (e.g., 100,000 square feet), but practically none for smaller offices (e.g., 15,000 square feet) [67]; the cost of a new system to provide office-

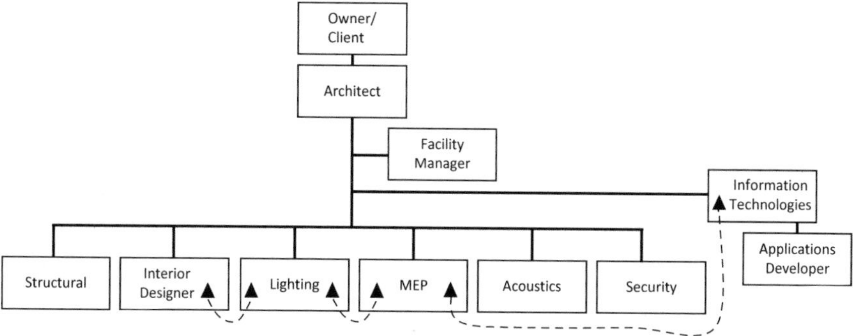

Fig. 3.23 Lighting project design/implementation team (high-level view)

quality illumination to a typical office environment is around \$25–30 per square foot. The advantages of connected light systems, besides the energy saving, is not in CAPEX reductions per se, but in the increased management flexibility of an integrated IoT/BMS-based environment. Additionally, there will be lower energy costs due to occupancy/ vacancy sensors and daylight harvesting. These sensors could, in theory, also be used to monitor employee spacing in the "new normal" office environment that is expected to follow the 2019/2020 pandemic.

A lighting project needs a complex design/implementation team to be successfully deployed. As seen in Fig. 3.23, for new construction, there is the site owner, client, and user team that will broadly define the requirements in conjunction with facilities management (and possibly the IT department); the design/implementation team includes, multiple disciplines prominently the architect, a lighting designer/consultant and other subject matter experts, notably an IT consultant, electrical engineer and possibly a security consultant—the lighting vendor may also be in the loop, as could be application/IoT developers (if enhanced sensing capabilities are required). For retrofit projects, where typically only the fixtures are replaced with no rewiring or new controls, a far less complex project team will be required with minimally the lighting designer, vendor and electrical contractor.

Besides ascertaining that the appropriate set of fixtures are included in the design, based on the requirements of the client and target environment, a technical design at the illumination level is needed. This is the province of the lighting designer. (In some cases there could be one-for-one replacements, such that no lighting designer is obligatorily required). Figure 3.24 provides a simplified example of such a design.

The system selection is important, also considering if the project is green field or a retrofit. Besides meeting the lighting requirements (from a functional—e.g., how many lumens are needed and where on the floor—aesthetic, and regulatory perspective), determination of the nature of the deployment needs to be made: will connected lighting be more traditional (with fixture-based or space-based sensors), or will it entail the more encompassing integrated IoT design? Will the system be

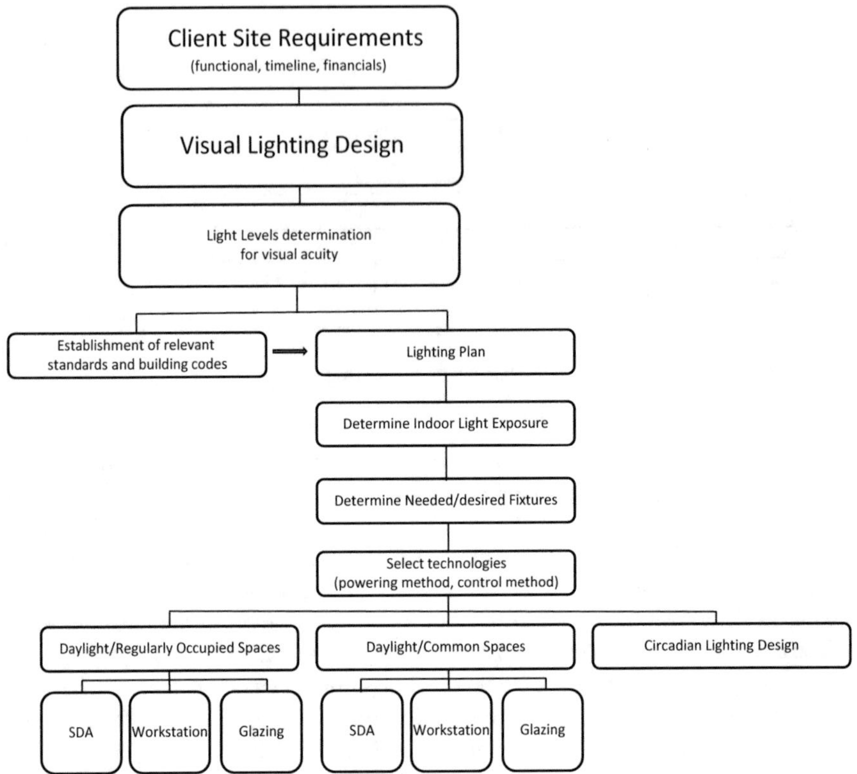

Fig. 3.24 Lighting design process for offices and other workspaces

traditionally-powered or will low voltage/DC solutions be sought? What panoply of fixtures with what panoply of sensors will be needed?

When deploying PoE LV systems Emergency PoE battery operated fixtures need to be developed or UPS system must be provided in IT rooms; also, Emergency PoE LV wiring must be run separately from normal PoE LV wiring [54].

3.9 Other Considerations

There are numerous additional practical, operational and technical considerations involved in the adoption and assimilation of any new technology into standard operating procedures, smart lighting being merely one example. A handful of these important issues are covered in this section.

Clearly industry regulations (stimulated in no small part by energy conservation initiatives) will accelerate (and /or force) the introduction of these systems in addition to continued price reduction in some of the core components. The approval of Authorities having Jurisdiction (AHJs) into local building codes will also enable many of these technologies to be selected as alternatives at the design stage to currently accepted methods. Additionally, success stories in the industry of working installations with demonstrable Return on Investment (RoI) over the life cycle of operation will spur wider acceptance. Conversely, a few failures will stymie efforts and require changes possibly to product, process and personnel.

Nevertheless, the trend towards connected lighting with wired and /or wireless controls is inexorable. As of this writing, several operational concerns have emerged that will ultimately need to be addressed.

The cost of deploying smart lighting may outweigh benefits in certain installations and only be adopted in spaces as dictated by code. For example, there are many types of light fixtures that require double and triple the maximum rated output of PoE port output, thus requiring multiple cable runs instead of a single line voltage cable. In addition, there are many instances (notably corridors, stairwells, etc.) where a single daisy-chained 20 A circuit can support many luminaires compared to multiple PoE runs. In such instances, the installation cost savings may be negligible or non- existent.

Day-to-day responsibility for the operation of lighting systems, which is the traditional province of the facilities department, may now have to be shared with the IT department if the lighting controls share a Wi-Fi or PoE infrastructure. This could lead to inertia to adoption if the partition of responsibility is not clear. Recent examples range from Facilities' understandable reluctance to share responsibility with IT (on the one hand) to IT's unwillingness to undertake additional responsibilities for day-to-day operation beyond their core mission (on the other). In addition, IT may outright refuse to allow yet another set of devices (not properly vetted from a cybersecurity perspective) to live on corporate networks. In the security space, certain cameras have been blacklisted for use (from some countries of origin because of the presence of malware). It is not inconceivable to envision similar bans being instituted for lights (manufactured overseas).

The maintenance of lighting systems, which typically entails troubleshooting from an electrical closet to the fixture, may now require also a trip to the IT/telephone closet where PoE switches or wireless access controllers are housed along with edge computing servers, raising additional security concerns and the perceived need for compartmentalization.

Additional obstacles to PoE adoption include the reality that no jurisdiction yet has approved deployment of emergency lighting systems on PoE, requiring line voltage to drive the emergency lighting systems. This aspect is decidedly a work in progress.

Some key concerns of smart lighting deal with cybersecurity. CLSs can be vulnerable to cybersecurity exploits that may be used by malicious hackers to gain access to a CLS-supporting network and compromise and misappropriate critical data. Downtime and ensuing loss of productivity in a commercial or industrial environment, significant harm to lighting and related assets, and reputational damage to the organization, are just a few of the risks.

In existing commercially available systems, and as discussed earlier, CLS devices are typically connected via a wireless gateway, via a direct wireless link, or via a wired switch; management is typically undertaken using web-based applications, accessed via on-premise server or a via a cloud-resident server. The availability of a large pool of Operation Technology (OT) data in CLSs, particularly network-based accessibility of that data, gives rise to cybersecurity risks and concerns that are generally unfamiliar and unusual to the lighting industry. These risks must be identified and addressed if CLS approaches are to be widely implemented.

In particular, authentication practices and their implementation are critical, especially when sensing and management are done over an intranet based Wi-Fi and/or PoE (IEEE 802.3) approach, even when relaying a local web server, since nearly all corporate and industrial intranets are connected to the Internet; the issue is accentuated in a cloud-based approach. These concerns also apply for CLSs using Zigbee based Mesh (IEEE 802.15.4), Bluetooth Mesh (Bluetooth Low Energy), or cellular (including 2G/GPRS and other cellular technologies). Some press time commercial CLSs did not implement integral authentication mechanisms and relied instead on mechanisms implemented for the web server. Recently the Pacific Northwest National Laboratory (PNNL), in cooperation with Underwriters Laboratories (UL), and under the auspices of the U.S. DoE, used their Connected Lighting Test Bed to conduct a series of studies to assess cybersecurity and risks in leading CLSs. The Laboratory concluded that the CLSs reaching the market had varying levels of cybersecurity risks and vulnerabilities; as a result, PNNL plans to conduct more authentication and related security testing [68, 69].

While most offices tend to have comparable lighting requirements, industrial settings are more diverse hence lighting requirements will be specific to the environment at hand, for example manufacturing and warehousing, food and beverage processing, and heavy industry. Industrial lighting requirements may depend on manufacturing cycles and the specific industrial sector. In some instances, multiple shifts are implemented, requiring near-continuous lighting. Manufacturing environments may require higher lighting intensity for productivity or safety reasons. Many factories have higher ceilings compared to office environments, as high as 30 meters, therefore the lighting design will have to be different than is the case for office environments.

There are two types of factory-related illumination, vertical and horizontal: vertical illumination is the light that addresses vertical surfaces, for example signs on a wall or labels on vertical storage bins; horizontal illumination is the light that addresses horizontal surfaces, for example on assembly lines, conveyor belts, or work benches; both of these requirements need to be met. Flicker should be minimized; glare, both direct or reflected, should be minimized—there are practical methods to address this issue, for example keeping objects that can create glare out of view. Industrial lighting systems also need to have high durability to function effectively in harsh conditions, such as high temperatures, high humidity, dust and vibrations, and possibly "dirty power". Additionally, these systems must comply with applicable codes related to safety, fire, green mandates for energy efficiency, and healthy work environments. Another expectation is that the luminaires should be reliable and require only low maintenance [70].

Organizations such as IESNA and OSHA recommend minimum lighting levels for performing specific factory tasks factored on the frequency of activity and the potential danger present. The IESNA Lighting Handbook offers the following guidelines as recommended lighting kevels (lux, lumens per square meter) [71, 72]: routine working spaces require a maintained illuminance target of 100 lux and lamps with a CRI of 70–85 or greater should be utilized to achieve adequate comfort levels; support of visual factory tasks requiring visual acuity for high contrast or large scale activities: 200–500 lux; support of visual factory tasks requiring visual acuity for medium contrast or small size: 500–1000 lux; support of visual factory tasks requiring visual acuity for low contrast or very small size: 1000–2000 lux; assembly/inspection that is very exacting: 3000 lux; and warehousing and storage dealing with bulky items, large labels: 300 lux. Many factory use robotics for some major portion of the manufacturing process, therefore the lighting requirements will also be different. In some cases, "lights out" operations are the norm. Warehouses will have different requirements from factories. Also, the scale of a factory or industrial environment may well preclude the use of PoE due to distance limitations. This topic deserves a separate discussion.

3.10 Conclusion

As discussed in this chapter, LED-based illumination requires two-to-three times less energy and requires up to 90% less maintenance compared with traditional lighting systems. Connected lighting systems further benefit from the fine-grain control that can be achieved. For example, dimming and light modulation capabilities of these connected systems allow for a more efficient utilization of electricity since lighting can be made adaptive to the environment, using occupancy-based sensors, daylight harvesting and Demand Response processes, thus further optimizing electricity expenditures incurred by the firm or organization. Enhanced controls, whether wired or wireless, can also provide facilities managers with real time usage and performance data. The lighting advancements discussed in this chapter support the sustainability, energy efficiency, and cost reduction goals being established worldwide. There is an expectation that the office environment may change in the future as a result of the 2019/2020 pandemic: in many industries, more employees will work from home; and, there will be a need to maintain safe distancing in the office. These factors may require more flexible seating density/ arrangements which would in turn influence the lighting requirements in the office while, at the same time, offering the possibility that CLSs can be used to monitor safe distancing of employees. Practitioners should continue to research the business benefits that these CLS technologies afford and endeavor to undertake appropriate deployments in smart cities, smart streets, smart buildings, smart factories, and smart homes, also assisted by the continued advancements in IoT principles and technologies.

References

1. Van Hoof, B.: Data-driven work spaces—IoT and AI expand the promise of smart buildings, pulse survey, Harvard Business Review Analytic Services, 24 Sept 2018. https://hbr.org/sponsored/2018/09/data-driven-work-spaces
2. The city of New York, Mayor Bill De Blasio, OneNYC 2050, building a strong and fair city, volume 1, Apr 2019. https://onenyc.cityofnewyork.us/wp-content/uploads/2019/05/OneNYC-2050-Full-Report.pdf
3. Wang, J.: Zigbee light link and its applications. IEEE Wirel. Commun. **20**(4), 6 (2013). https://doi.org/10.1109/MWC.2013.6590043
4. U.S. Energy Information Administration: Commercial buildings energy consumption survey (CBECS), energy usage summary, 2012 edition. http://www.eia.gov/consumption/commercial/reports/2012/preliminary/
5. CBECS Status, 22 Apr 2019. https://www.eia.gov/consumption/commercial/
6. Energy.gov, Rise and shine: Lighting the world with 10 billion LED Bulbs, 7 Dec 2015
7. U.S. Energy Information Administration (EIA), as reported by NIH/NLM/TOXNET
8. Pandharipande, A., Newsham, G.R.: Lighting controls: evolution and revolution. Light. Res. Technol. **50**, 115–128 (2018)
9. Tan, F., Caicedo, D., et al.: Sensor-driven, human-in-the-loop lighting control. Light. Res. Technol. **50**, 660 (2017). https://doi.org/10.1177/1477153517693887
10. Energy.gov, Office of Energy Efficiency and Renewable Energy - Building Technologies Office, Next Generation Lighting Systems Launches Next Phase of Indoor Evaluations; Outdoor Evaluations Set to Start, Press Release August 16, 2019
11. IESNA, Illuminating Engineering Society of North America. https://www.ies.org/
12. Williamson, S., Cummins, H.: Light and color in nature and art. Wiley, New York (1983)
13. Minoli, D.: Imaging in corporate environments - technology and communication. McGraw-Hill, New York (1994)
14. ANSI/NEMA spec C78.377.2008: Specification for the chromaticity of solid state products
15. ANSI/NEMA spec C82.77-2002: Harmonic emission limits - related power quality requirements for lighting
16. IESNA spec TM-16-05: IESNA technical memorandum on LED sources and systems
17. IESNA spec LM-80-08: IESNA approved method for measuring lumen maintenance of LED light sources
18. Staff, the International WELL Building Institute, Circadian Lighting Design. https://standard.wellcertified.com/light/circadian-lighting-design
19. ERCO guide. 2010. www.erco.com
20. Hassan, Q. (ed.): Internet of things A to Z: technologies and applications. Wiley, Hoboken (2018)., ISBN-13: 978-1119456742
21. Hassan, Q., Khan, A.R., Madani, S.A. (eds.): Internet of things: challenges, advances and applications. CRC Press, Taylor & Francis, Boca Raton (2018)., ISBN 9781498778510
22. Dehghantanha, A., Choo, K.-K.R. (eds.): Handbook of big data and IoT security. Springer, Cham (2018, ISBN 978-3-030-10542-6). https://doi.org/10.1007/978-3-030-10543-3
23. Aazam, M., Huh, E.-N.: Fog computing and smart gateway based communication for cloud of things. In: Proceedings 2nd International Conference on Future Internet of Things and Cloud, FiCloud-2014, Barcelona, Spain, Aug 2014
24. Al-Fuqaha, A., Guizani, M., et al.: Internet of things: a survey on enabling technologies, protocols, and applications. IEEE Commun. Surv. Tutorials. **17**(4), 2347 (2015)
25. Stergiou, C., Psannis, K.E., et al.: Secure integration of IoT and cloud computing. Futur. Gener. Comput. Syst. **78**, 964–975 (2018)
26. Minoli, D., Occhiogrosso, B., et al.: Multimedia IoT systems and applications, global IoT summit, GIoTS-2017, Organized by Mandat International, IEEE IoT TsC (Transactions on Service Computing), the IoT Forum and IPv6 Forum (collocated with the IoT Week), 6–9 June 2017, Geneva

27. Minoli, D., Occhiogrosso, B.: Ultrawideband (UWB) technology for smart cities IoT applications. In: 2018 IEEE International Smart Cities Conference (ISC2) - IEEE ISC2 2018-Buildings, Infrastructure, Environment Track, Kansas City, 16–19 Sept 2018
28. Reyna, A., Martín, C., et al.: On blockchain and its integration with IoT. Challenges and opportunities. Futur. Gener. Comput. Syst. **88**, 173–190 (2018). https://doi.org/10.1016/j.future.2018.05.046
29. Minoli, D., Occhiogrosso, B.: IoT applications to smart campuses and a case study. EAI Endorsed Trans. Smart Cities. **2** (2017). https://doi.org/10.4108/eai.19-12-2017.153483
30. Biery, E., Shearer, T., et al.: Controlling LEDs, technical white chapter, May 2014, Lutron. http://www.lutron.com/TechnicalDocumentLibrary/367-2035_LED_white_chapter.pdf
31. Staff, Sololuce, LED lighting information, Sept 2013. www.sololucegroup.com/Sololuce%20technical%20Knowledge.pdf
32. J. Room, 5 Charts that illustrate the remarkable LED lighting revolution, Aug 2, 2016. https://thinkprogress.org/5-charts-that-illustrate-the-remarkable-led-lighting-revolution-83ecb6c1f472/
33. U.S. DoE: Energy savings forecast of solid-state lighting in general illumination applications, 2016. https://www.energy.gov/eere/ssl/ssl-forecast-report
34. IHS MARKIT Press Release, 21 Dec 2017. www.businesswire.com/news/home/20171221005630/en/; www.ihsmarkit.com
35. Higuera, J., Llenas A.: Trends in smart lighting for the Internet of Things, 29 Aug 2018, Cornell University, Computers and Society (cs.CY), arXiv:1809.00986 [cs.CY]
36. Feris, M.: Session 0216: Why 'compatibility' is the magic word in controlling LEDs? Designers Light Forum, March 2018. leducation.org
37. Qin, Z., Sun, Y., et al.: Enhancing efficient link performance in ZigBee under cross-technology interference. Mobile Netw. Appl. **25**, 68 (2019)
38. Yi, P., Iwayemi, A., Zhou, C.: Developing Zigbee deployment guideline under Wi-Fi interference for smart grid applications. IEEE Trans. Smart Grid. **2**(1), 110 (2011). https://doi.org/10.1109/TSG.2010.2091655
39. ZigBee Light Link Standard Version 1.0, ZigBee Document 11–0037–10, Apr. 2012
40. Yang, J., Liu, R., Cui, B.: Enhanced secure ZigBee light link protocol based on network key update mechanism. In: International Conference on Innovative Mobile and Internet Services in Ubiquitous Computing (IMIS) 2018, Innovative Mobile and Internet Services in Ubiquitous Computing, Part of Springer Advances in Intelligent Systems and Computing book series (AISC, volume 773) pp. 343–353
41. Yang, J., Liu, R., Cui, B.: Improved secure ZigBee light link touchlink commissioning protocol design. In: 2018 32nd International Conference on Advanced Information Networking and Applications Workshops (WAINA), Krakow, Poland, 16-18 May 2018. https://doi.org/10.1109/WAINA.2018.00138
42. Olawumi, O., Haataja, K., et al.: Three practical attacks against ZigBee security: attack scenario definitions practical experiments countermeasures and lessons learned. In: IEEE 14th International Conference on Hybrid Intelligent Systems (HIS2014) at Kuwai
43. Woolley, M.: How Bluetooth mesh puts the 'large' in large-scale wireless device networks, 26 June 2018. https://www.bluetooth.com/blog/mesh-in-large-scale-networks/
44. M-Way Solutions Staff, Smart Beacon Management with BlueRange, Version 1.1 – Status 01/2018, M-Way Solutions GmbH, Stresemannstr. 79, 70191 Stuttgart, Deutschland. https://www.bluerange.io/downloads/en_bluerange_guide_2018.pdf
45. Slupik, S.: Bluetooth mesh networking: the packet, 2017. https://www.bluetooth.com/blog/bluetooth-mesh-networking-the-packet/
46. Minoli, D., Sohraby, K., Occhiogrosso, B., et al.: IEEE Internet Things. **4**(1), 269–283 (2017). https://doi.org/10.1109/JIOT.2017.2647881
47. Price, C.: Lighting: pros and cons of using DALI with KNX for homes, 6 June 2019. http://knxtoday.com/2019/06/13755/lighting-pros-and-cons-of-using-dali-with-knx-for-homes.html
48. DALI. www.dali-ag.org

49. Philips Advance Staff, The ABC's of DALI: A guide to digital addressable lighting interface, 2009, document CO-7110-R03 11/09, Philips Lighting Electronics N.A., 10275 W. Higgins Road, Rosemont, IL 60018
50. Mesh, S.: Power over ethernet lighting systems, 11/29/2017, Lighting Controls Association. https://lightingcontrolsassociation.org/2017/11/29/steve-mesh-on-power-over-ethernet-lighting-systems/
51. International WELL Building Institute. https://www.wellcertified.com/about-iwbi/
52. DesignLights Consortium (DLC). https://www.designlights.org/about-us/
53. Farris, P., Lehman, M.: Deciphering IECC, ASHRAE 90.1, and title 24, part 6 lighting and lighting control requirements, designers light forum. Leducation.org
54. Widmer, J., Ziegenbein, P., Zinkon, A. P.: IoT connected lighting: a design guide, designers light forum, 12 Mar 2019. Leducation.org
55. Parks, B.: Community friendly lighting 101, designers light forum, 12 Mar 2019. Leducation.org
56. IES, ANSI/IES RP-8-18: Recommended practice for design and maintenance of roadway and parking facility lighting, 2018
57. Market Research Future, Smart Lighting Market 2019 Global Applications, Industry Size, Development Status, Regional Analysis, Competitive Landscape and Recent Trends by Forecast to 2027, 14 May 2019. https://www.marketresearchfuture.com. Also https://bestmarketherald.com/smart-lighting-market-2019-global-applications-industry-size-development-status-regional-analysis-competitive-landscape-and-recent-trends-by-forecast-to-2027/
58. P&S Intelligence, Global-Smart-Lighting-Market-is-Anticipated-to-Reach-US-27-064-million-by-2024, Press Release, 05-20-2019. https://www.openpr.com/news/1745609/Global-Smart-Lighting-Market-is-Anticipated-to-Reach-US-27-064-million-by-2024-Having-Major-Players-Cree-Hubbell-Lighting-Honeywell-Acuity-Brands-Philips-Lighting-OSRAM-Schneider-Electric-Legrand.html
59. Global smart lighting market analysis, smart lighting market expected to witness a sustainable growth over 2025 – top key players: Acuity Brands, Inc., Zumtobel AG, Digital Lumens, Inc., Streetlight.Vision, Encelium Technologies, Press release from: Worldwide Market Reports, 05-15-2019. www.worldwidemarketreports.com
60. Databridge Market Research, Global Lighting Control System Market 2024 Outlook Worth, Size, Growth, Trends & Industry Analysis by Leviton Manufacturing Company, Inc., Echelon Corporation, Lightwaverf Plc, Cree Inc., Lutron Electronics Co. Inc., 20 May 2019. https://databridgemarketresearch.com/request-a-sample/?dbmr=global-lighting-control-system-market
61. Reports and Data Staff, Power Over Ethernet (POE) Lighting Market To Reach USD 1.27 Billion By 2026. https://www.globenewswire.com/news-release/2019/04/02/1795631/0/en/Power-Over-Ethernet-POE-Lighting-Market-To-Reach-USD-1-27-Billion-By-2026.html. Also https://www.reportsanddata.com/report-detail/power-over-ethernet-poe-lighting-market
62. Adroit Market Research, Power over Ethernet (PoE) Lighting Market to Hit $13.50 Billion by 2025 - Global Analysis by Size, Share, Trends, Key Players & Opportunities, 16 April 2019. https://www.globenewswire.com/news-release/2019/04/16/1804714/0/en/Power-over-Ethernet-PoE-Lighting-Market-to-Hit-13-50-Billion-by-2025-Global-Analysis-by-Size-Share-Trends-Key-Players-Opportunities-Adroit-Market-Research.html
63. Schweber, B.: PoE + LEDs: an effective pairing, with some limitations. https://www.avnet.com/wps/portal/us/resources/technical-articles/article/markets/lighting/poe-leds-an-effective-pairing-with-some-limitations/
64. DOE Office of Energy Efficiency and Renewable Energy, Connected Lighting Systems Efficiency Study—PoE Cable Energy Losses, Part 1, November 2017 (revised January 2019)
65. ANSI C137.3-2017, American National Standard for Lighting Systems – Minimum Requirements for Installation of Energy Efficient Power over Ethernet (PoE) Lighting Systems, Secretariat: National Electrical Manufacturers Association. Approved: May 25, 2017, American National Standards Institute, Inc.

66. DOE Office of Energy Efficiency and Renewable Energy, PoE Connected Lighting System Energy Losses in Ethernet Cables – Part 2, 19 Dec 2018. https://www.energy.gov/eere/ssl/articles/poe-connected-lighting-system-energy-losses-ethernet-cables-part-2
67. Pirot, J., Zeccardi, R., et al.: The era of smart buildings utilizing power over Ethernet, designers lighting forum, March 2018. leducation.org
68. Energy.gov Offices, Energy Efficiency & Renewable Energy, Connected Lighting Cybersecurity Testing. https://www.energy.gov/eere/ssl/connected-lighting-cybersecurity-testing. Accessed 2 Nov 2019
69. Poplawski, M., St. Lawrence, A.: Stress test cybersecurity lessons emerge from a recent study of connected lighting, LD+A, June 2019, Illuminating Engineering Society. https://www.energy.gov/sites/prod/files/2019/06/f63/connectedlighting_lda_june2019.pdf
70. Resiliant Staff: Key considerations for lighting industrial facilities, May 2015. http://lumefficient.com/wp-content/uploads/2018/09/industrial-key-considerations.pdf
71. DiLaura, D., Houser, K., et al.: The lighting handbook: 10th edition, reference and application. Illuminating Engineering Society of North America, London (2011). ISBN-13: 978-0-87995-241-9
72. Saif Staff: Safety topic, industrial hygiene, lighting for office and industry, document SS-405. Saif.com/safetyandhealth

Chapter 4
Automation Trends in Industrial Networks and IIoT

David Camacho Castillón, Jorge Chavero Martín, Damaso Perez-Moneo Suarez, Álvaro Raimúndez Martínez, and Victor López Álvarez

4.1 Introduction

At the beginning of the century, the changes on business model forced a new industrial revolution. The motivation of this Fourth Industrial Revolution are cost reduction, rapid technology adoption, end product customization, and product delivery time reduction. New breakthroughs in technology support this new productive model. Those enabling technologies allow the new productive model implementation. Each industry will only implement the technologies that will enable new production requirements. This approach impacts on the traditional automation practices in factories [22].

Industrial manufacturing and production environments have been gradually adding to their systems the support of Information Technology (IT) advantages. IT systems aim to control technological solutions as well as the logistic processes, which are becoming more and more complex. The information exchange between the factory and the corporate system increases performance and flexibility in the production process. Currently, factories' evolution toward digitalization, together with enabling technologies, allow the new productive models implementation.

D. Camacho Castillón (✉) · J. Chavero Martín · D. Perez-Moneo Suarez · Á. Raimúndez Martínez
GEA, Alcobendas, Madrid, Spain
e-mail: david.camacho@gea.com; jorge.chavero@gea.com; damaso.perez-moneo@gea.com; alvaro.raimundez@gea.com

V. L. Álvarez
Telefónica I+D, Ronda de la Comunicación S/N Madrid, Madrid, Spain
e-mail: victor.lopezalvarez@telefonica.com

I. Butun (ed.), *Industrial IoT*, https://doi.org/10.1007/978-3-030-42500-5_4

4.2 Industrial Revolutions: From The First Industrial Revolution to Industry 4.0

To better understand Industry 4.0, the following section shows the most relevant items of the four industrial revolutions that have led human beings to the social, economic, and industrial development that we have nowadays.

4.2.1 First Industrial Revolution: Mechanization Revolution and Steam-Powered Engine

The First Industrial Revolution, also known as the industrial revolution or Industry 1.0, dates from the 1780s to the 1830s. It began in Great Britain, where the most significant changes occurred in the field of production processes, transforming an industry based on manual operations to activities carried out by machines [19].

A notable development of this period was the use of steam as the primary source of thermal energy to generate mechanical power. Devices such as the James Watts machine, patented in 1769, or the Thomas Newcomen steam machine were precursors of the steam train, that was initially used to transport merchandise and later people over long distances. The evolution of these steam engines, with an unprecedented power magnitude allowed growth in mining and mineral extraction, fostering the growth of sectors such as the chemical industry, with the industrial production of sulfuric acid, the first time developed by John Roebuck in 1746 or the cement sector. The emerging exploitation of coal and the development of blast furnace for coke had an essential role in the increase of steel production at a low cost, improving growth in machine building and the civil construction as bridges or iron rails [7, 25].

The textile industry underwent a significant transformation due to the integration of new spinning machines that raised production levels, reducing processing time from a few months to days. In 1733, "The flying shuttle" was patented by John Kay, and after numerous improvements, in 1789, Edmund Cartwright invented the "vertical power loom." Fields such as Cotton Industry notably benefited from inventions such as "Water Frame" by Richard Arkwright's in 1769 or "Spinning Mule" 1779 by Samuel Crompton's leading to high-quality production with minimum labor [19].

4.2.2 Second Industrial Revolution: Electricity and Mass Production

The Second Industrial Revolution or Industry 2.0 covers from 1870 to approximately 1914. The widespread use of electricity integrated into society and factories through an electrical grid, thanks to the contributions of inventors as Thomas

Alva Edison or Nikola Tesla, characterizes this revolution. The electrical power generation from hydraulic resources favors a rapid proliferation of industrial areas, and electric light began to replace the light coming from the combustion of gas. Electrical engines substituted the steam engines developed during the first industrial revolution [14].

This period witnessed the beginning of the use of oil as a source of raw material and energy suppliers. The development of internal combustion engines enabled the construction of the first vehicles and means of transports that remain today.

The industrial mass production in assembly lines came from the hand of Henry Ford. The introduction of the Ford T in the automobile market in the 1900s revolutionized transportation and industry in the United States. In 1911, Frederic Winslow Taylor published "The principles of Scientific Management," where he put forward that the simplification of tasks and their optimization imply increasing production, setting the precedents of the current lean manufacturing and management methods [28].

4.2.3 Third Industrial Revolution: Digitalization and Automation

The Third Industrial Revolution or Industry 3.0 covers the second half of the twentieth Century until the 2000s. The development of electronics and information technologies represents this revolution. After the invention of the transistor in 1947, John Bardeen, Walter Houser Brattain, and William Bradford Shockley received the Nobel Prize in Physics 1956 for their researches on semiconductors and their discovery of the transistor effect. This revolutionary item introduced a new era of computing and communications, facilitating the creation of innovative devices as the first Personal Computer (PC), and Programmable Logic Controller (PLC) [27].

The first PLC, Modicon, was designed in 1969 by Bedford Associates, requested by General Motors. Dick Morley is considered the inventor of the PLC due to his role in its development and implementation. The PLC is the principal agent of the industrial automation of processes and replaced the previous hard-wired systems, relays, timers, and sequencers, improving methods in terms of product quality and process complexity [12].

Since the invention and flew off the first airplane of the Wright brothers, in 1903, the commercial aviation industry is driven towards continuous improvement of aircraft, making them safer and meeting the needs of a globalized economy [26].

In 1942, Enrico Fermi achieved the first controlled nuclear chain reaction framed in The Manhattan Project (1942–1946), confirming Albert Einstein's publication of 1905, mentioning the equivalence between mass and energy. Thermonuclear energy was commercial with the nuclear fission reaction of Uranium-235 in the 1960s by Westinghouse and General Electric companies in The United States [1].

Brenners-Lee founded the World Wide Web Consortium in 1994, intending to supervise and standardize the development of technologies that allow the operation of the Internet. In the 2000s, the use of the Internet is global, and its integration in devices such as mobile phones has had a high penetration [24].

4.2.4 Industry 4.0

The Fourth Industrial Revolution or Industry 4.0 begins in the 2010s. This revolution describes the globalization of digitalization transformation and automation trends supported by technologies such as the Industrial Internet of things (IIoT), Cyber-Physical Systems (CPS), Cloud Computing, Big Data Analytics, 5G networks, Cybersecurity, Augmented Reality (AR), among others. Industry 4.0 introduces Smart manufacturing or Smart Factory concepts where devices and systems acquire higher intelligence. Organizational and internal services manage the information, and both Machine to Machine (M2M) and Human to Machine real-time communications, providing excellent value to companies, organizations, and society.

Many actors work in collaboration to drive this industrial revolution. Politicians, industries, companies, universities, the academic scientific sector, associations, and trade unions work together to standardize, promote knowledge, set guidelines, and implement the set of technologies of this Fourth Industrial Revolution. The two main projects that lead this fourth industrial revolution are the Plattform Industrie 4.0 and The Industrial Internet Consortium (IIC) [23].

The Plattform Industrie 4.0 was publicly announced at the Hannover Fair in 2013, framed in the "2020 High Technology Strategy Action Plan" of the Federal Republic of Germany. Industrie 4.0 promotes digitalization for small and medium enterprises (SMEs). It coordinates information and networking services across Germany and other countries. Its technical approach is next-generation manufacturing value chain focusing on the supply chain, embedded systems, automation, and robotics [23].

The IIC, founded in March 2014, is a non-profit world's leading membership program of the Object Management Group (OMG). It aims to transform society and businesses through the acceleration of the Internet of Things. Among the members are innovative companies, large multinationals, market leaders, researchers, universities, and government organizations. The emphasized founders are General Electric, DELL, Microsoft, Huawei, or Bosch. IIC apply to IIoT across energy, healthcare, manufacturing, public domain, and transportation industries, overall [9, 15].

The following section shows in detail the leading enabling technologies for this fourth industrial revolution, mainly focused on the manufacturing industry.

4.3 Enabling Technologies for New Productive Model

4.3.1 IoT and Big Data

Kevin Ashton was the first to introduced the term "Internet of Things" to refer to the elements/sensors which are connected to the internet. Kevin predicted that these devices would outnumber human beings. Nowadays, that prediction is correct [8, 30]. The increase of elements connected to the internet multiplies the information accessible to the companies, which can have more data about potential buyers and its factories. With the proper selection, processing, and research of that data, companies can improve their performance. Industrial devices produce in real-time more quantity and accurate data every day about the machine and its environment both. The correct understanding of these data can provide the factory managers with tools to improve the production schedule, machine maintenance, and avoid possible faults and time inefficiencies. "You can't manage what you don't measure" [4], or in other words, once we have access to more information, we can improve the functionality and performance of the plants. As Jay Lee, Hung-An Kao, Shanhu Yang indicate [13] the right use of that significant amount of data, for a factory, can be split into five areas:

- The *Manager and Operator interaction:* It is possible to improve the production schedule and get the maximum performance of the machines.
- The *Machine fleet:* Strongly related to the previous point, it is possible to share the use of the devices and then reduce the use, repeatability, and the wear of the tools, extending their useful life.
- The *Product and Process Quality:* Quality and final product cost is the most critical topic in a factory. If we can relate the ultimate quality and price of the product with all the previous processes related to it, we could improve the manufacture of the products on many different levels, being more competitive as a whole.
- The *Big Data and Cloud:* Saving, sharing, and analyzing all these data is vital for the future.
- The *Sensors and Controller Networks:* New sensors provide more information about themselves, regarding their use, deterioration, and maintenance requirements. Besides, they could give information about its environment and the machine.

In essence, now we can have wider access to the elements and sensors, receiving a massive amount of information that we need to learn how to process and use. If we could take advantage of all those data, we could improve the production schedule and machine performance. As Google's Chief Scientist Peter Norvig admitted: "We don't have better algorithms than anyone else; we just have more data." [6].

4.3.2 Cloud Computing

As it was mentioned in the previous point about IIoT and Big Data, with the increasing digitization of industrial sensors and actuators, the number of communications and the amount of data are growing every day. Thus, the need to make the right use of them promotes the emerging of new applications such as Big Data Analytics, IIoT, Artificial Intelligence (AI), Machine Learning, among others, as described during the chapter. These applications and enabling technologies require higher processing power, storage capabilities, and new staff skills to host, use, and maintain them, that the traditional OT infrastructures don't have.

The first time the term "Cloud Computing" became popular was around 2006 when large companies as Amazon or Google offered a kind of cloud service. It is now with the integration of industrial systems with corporate systems and the Internet that the use of this enabling technology is proposed in its different modalities of use [17] (as shown in Fig. 4.1):

- The *SaaS (Sofware as a Service):* Allows users to connect to and use cloud-based apps over the Internet.
- The *PaaS (Platform as a Service):* Complete development and deployment environment in the cloud, with resources that enable to deliver everything from simple cloud-based apps to sophisticated, cloud-enabled enterprise applications.
- The *IaaS (Infrastructure as a service):* Computing infrastructure, provisioned, and managed over the internet.

Even though there are well known Cloud Computing advantages as hardware independency, cybersecurity, scalability, harmonization, low maintenance, higher implementation times with less risk. However, it also has drawbacks. Some disadvantages go against the industrial sector's fundamental requirements: availability, response time, or privacy.

Because of the above, on-premise solutions are becoming very popular within the industrial sector because they fit the computing and storage requirements without an external dependency at a competitive price. It's called Edge or Fog computing and allows to manage a massive volume of data with low latency.

Not all data has business value, nor does it provide information. That is why it should be analyzed what data should be potentially stored in the Cloud. It's another application example for Edge computing that it's not coming to supply Cloud but to complement it.

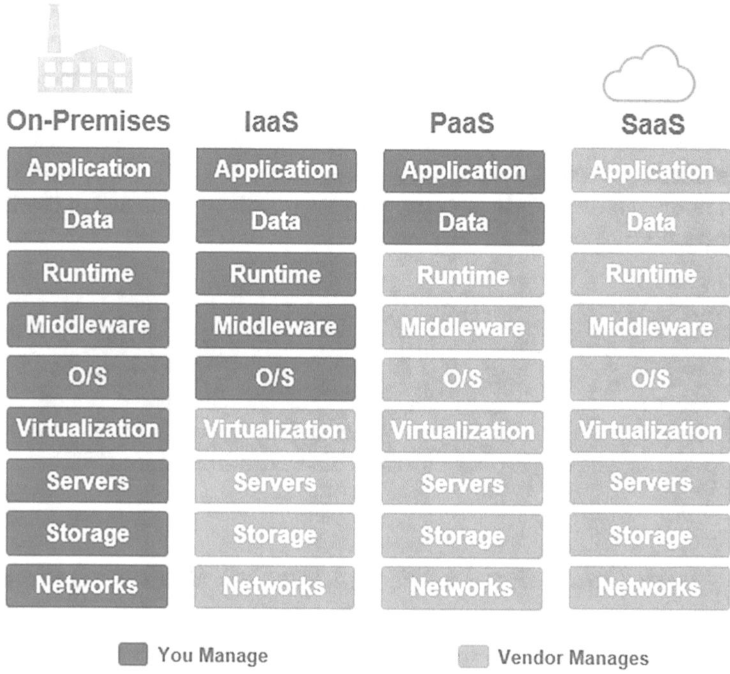

Fig. 4.1 Cloud computing service model

4.3.3 *Industrial Cybersecurity*

Since the emergence of computer and automation into the industrial scene (Industry 3.0), the divergences in characteristics and objectives between the Information Technology (IT) and the Operational Technology (OT) have led their technological evolution until now.

OT systems focus on availability and determinism over confidentiality through systems characterized by limited configuration, heterogeneity, long life cycles, and few software alternatives. Furthermore, the primary concern of the industrial sector has been the preservation of productive processes, making then so reluctant to changes that may cause downtimes. "If it ain't broke, don't fix it." Therefore, the OT systems aware of their shortcomings and sensitivity have been basing their security by using perimeter security, keeping these cells isolated from corporative networks, and even more from the Internet. Meanwhile, the IT systems were suffering a progressive and constant transformation by integrating Quantitative Risk Assessment (QRA), Security Operation Center (SOC), Manage detection and Response, penetration testing, audits, Incident Response Plan, and digital forensic.

It's only now, with the emergence of these disruptive technologies and new business models, that OT Systems have been forced to be integrated with the IT networks. The confluence with IT has been exposed to the immaturity level with regards to security caused by this isolation for more than 20 years.

Cybersecurity is a cultural change, which is not only a purely technological solution, but a complete framework implementation that manages the corporative risk assessment and presents protection measures. Security's not a state; it's a process (Identify, Protect, Detect, Respond, and Recover) [20]. These functions and the rest of the Industrial Cybersecurity framework will be explained in detail in the specific section below.

Industrial Cybersecurity is considered as a transversal enabling technology upon which the rest of the enabling technologies should be designed and developed in the way to the Industrial Digital Transformation, which will allow improving the performance without assuming any added risk.

4.3.4 Fifth Generation Mobile and Wireless Communications

The fifth generation (5G) of mobile and wireless communications is designed not just as a technological solution, but to have a significant impact on industry and society [16]. 5G comes with three sets of use cases: enhanced Mobile Broadband (eMBB), Ultra-reliable low-latency communication (URLLC), and massive Machine Type Communications (mMTC). The 3GPP defined them as part of its Study on New Services and Markets Technology Enablers (SMARTER) project.

eMBB covers data-driven use cases requiring high data rates across a wide coverage area. This use case requires higher capacity, enhanced connectivity, and higher user mobility. The first requirement of higher capacity is the first feature for each evolution in a new mobile and wireless generation. The idea is that 5G provides broadband access to densely populated areas not just outdoors (city centers), but also indoors (office buildings, stadiums, or conference centers). The second requirement is enhanced connectivity. This concept extends the service available to any location so that the user can have a consistent experience. Finally, higher user mobility will enable such services in moving vehicles, including cars, buses, trains, and planes.

URLLC use cases are conditioned by extreme delay requirements, reliability, and availability. 5G NR is the new radio access solution that redesigns how the terminals are connected to the network. Thanks to this technology for 5G, URLLC services will have latency bellow 1 ms. 5G capable terminals must have to synchronize to the reference clock, thus allowing the traffic shapers to be time-aware, providing this low latency. URLLC has packet delivery success probability as high as 5-nines (1-10-5) to 9-nines (1-10-9), improving the reliability and availability of previous services.

mMTC are services that have a vast number of devices, transmitting (typically) low volume of non-delay sensitive data. The challenge is to have a technology that enables low-cost devices with a very long battery time (10 years).

ITUR [10] defines the framework and objectives of the future development of International Mobile Telecommunications (IMT) for 2020 and beyond. Each service is determined based on eight critical capabilities of IMT-2020; they are:

- *Peak data rate* [Gbit/s]. Maximum achievable data rate under ideal conditions per user/device.
- *User experienced data rate* [Gbit/s]. The achievable data rate that is available across the coverage area to a mobile user/device.
- *Latency* [ms]. The contribution by the radio network to the time from when the source sends a packet to when the destination receives it.
- *Mobility* [km/h]. Maximum speed at which a defined QoS and seamless transfer between radio nodes can be achieved.
- *Connection density* [km2]. The total number of connected and or accessible devices per unit area.
- *Energy efficiency* [bit/Joule]. Energy efficiency has two aspects: the network and the terminal side.
- *Spectrum efficiency* [bit/s/Hz]. Average data throughput per unit of spectrum resource and cell.
- *Area traffic capacity* [Mbit/s/m2]. Total traffic throughput served per geographic area.

Figure 4.2 shows how these eight key parameters are related to the use case categories presented before [10]. As an example, URLLC services will be those that have high mobility and latency requirements and low requirements for the rest. On the other hand, mMTC service will be those that need a high connection density, medium network energy efficiency, and low requirements for the rest.

Figure 4.3 represents examples of services for each of the three main use case categories defined for 5G networks [10]. Industry automation is mapped as a URLLC services due to its requirements in terms of latency. Sensors and monitoring devices are within the category of mMTC, that are part of the industrial environment. Further, in this chapter, we will present which are the possible scenarios for industrial applications in 5G networks.

Even though 5G technologies is an extensive topic [16], we want to highlight a technological evolution concept provided by 5G networks: network slicing. Network slicing permits efficient resource sharing of a common infrastructure among different services [21]. The network slicing concept expects some independent slices, each comprising different network nodes and functions, which are interconnected for a specific service or user. The use of network slicing will allow network operators to provision and control various services (like eMBB, mMTC, or URLLC) that have different requirements using the same underlying physical infrastructure. At the same time, each network slice must be set up dynamically without impacting on services already provisions, which requires to coordinate with other orchestration entities.

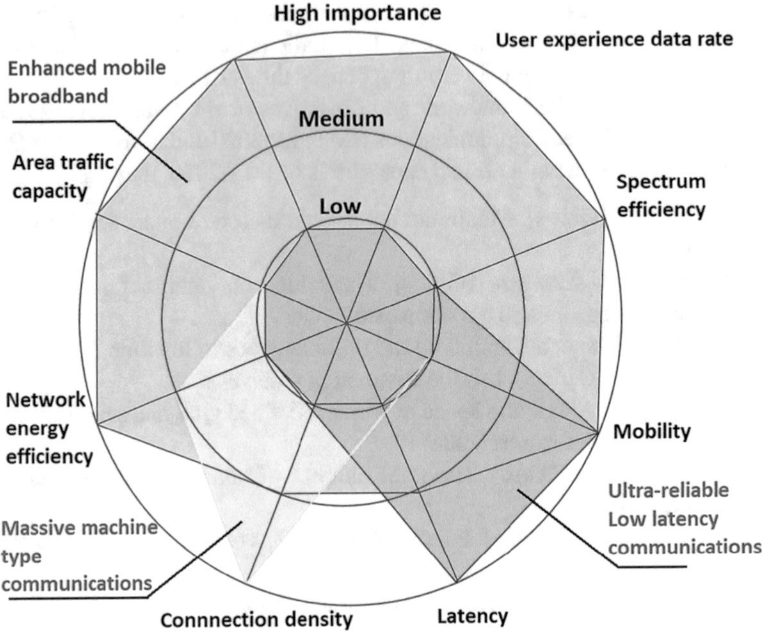

Fig. 4.2 Critical capabilities in different usage scenarios

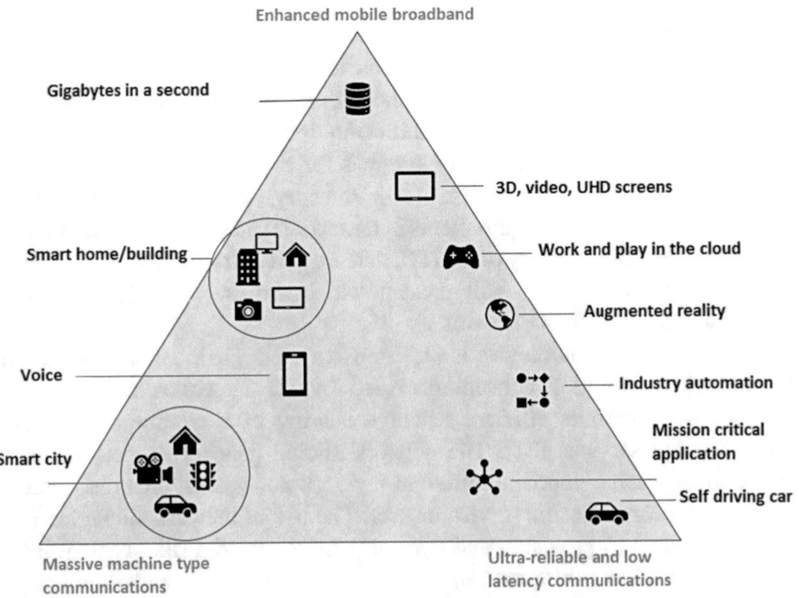

Fig. 4.3 Usage scenarios of IMT for 2020 and beyond

4.4 Automation Network in Smart Industries

The classic automation pyramid was defined by two standards: the Purdue Enterprise Reference Architecture and the ISA 95 during the third industrial revolution. Purdue Enterprise Reference Architecture (PERA), Theodore J. Williams, is a 1990s reference model for enterprise architecture, which describes the different levels. This division sets a shared understanding of an enterprise organization. PERA categorizes devices and a group of devices, based on physical, logical, and relational properties [2].

ISA-95 is an international standard from the 2000s that describes guidelines to develop an automated interface between enterprise systems and a control system. ISA-95 describes Level 3 functions and data interaction. This standard provides consistent terminology, information models, boundaries between systems, and what information could be exchanged between the different systems. Combined, they defined the Automation Pyramid (as shown in Fig. 4.4) [29]:

- *Level 0:* Include the physical process devices and installation. The devices on this layer are physical devices directly involved in the enterprise operation. Those devices do not have any connection to the upper system.
- *Level 1:* This level includes devices such as sensors, actuators, or analyzers connected to the upper layer. Those devices are connected through direct electrical connection or/and a communication port to a field network. Those devices control or analyze items in Level 0. This connectivity allows control, operation, and data exchange.

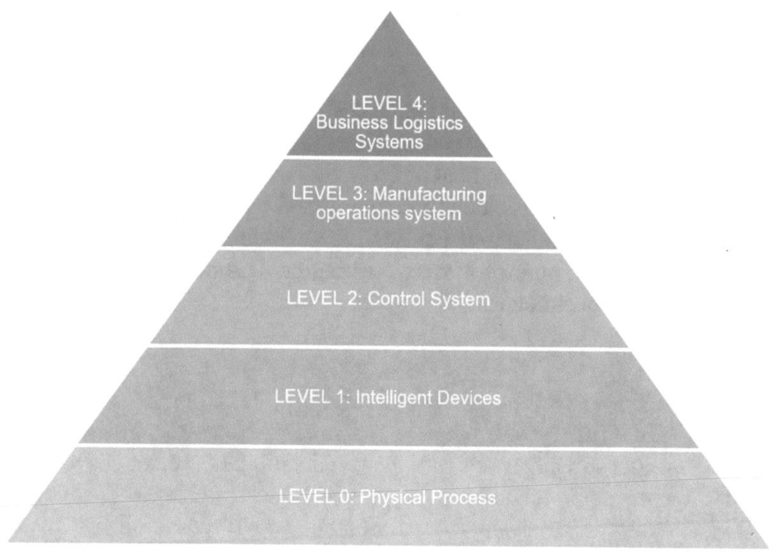

Fig. 4.4 Automation pyramid

- *Level 2:* Used to be divided into two sub-levels, monitoring, and Control:
  - *Control level:* Manages the Level 1 devices, bundling their information, and sending commands back to them according to predefined instructions coded on a PLC sequence or from an HMI/SCADA, answering a human request.
  - *The monitoring level:* Creates an interface between Level 1 and the humans who operate the system, representing the data, and allowing the interaction with the system. This level includes PLC/HMI, SCADA, and DCS systems.
- *Level 3:* ISA 95 focus. This layer manages the interaction between the enterprise systems (Level 4) and the control systems (Level 0, 1, and 2). Is is also referred to as MES (Manufacturing Execution System) or MOM (Manufacturing operations management). This layer translates real-time production data to transactional information.
- *Level 4:* Fundamentally, but not limited to, ERP systems. ERP system could interact with the lower levels, gathering information from the process, such as production quantity, and sending commands, like production orders.

Those layers shaped the control system designs during the last 30 years. The levels were designed following the functionality described, matching physical devices to tiers, and using different networks to communicate the layers. The classic architecture design split the networks accordingly:

Networks are the highway to transport information. The Classic control network architecture is based on Fieldbus networks. Those architectures usually include a Human Machine Interface (HMI) or a SCADA (Supervisory Control And Data Acquisition) on the top of level 2, where an operator can monitor or operate the system. HMI is linked to Programmable Logic Controllers (PLC). Communicating the level 2 and level 1, a network called Fieldbus connects PLCs to field components, such as sensors, actuators, electric motors, console lights, switches, valves, and contactors.

Fieldbus is the network that communicates Level 1 field devices with Level 2 Control System. The main Fieldbus characteristic is deterministic communication, required in an industrial environment due to safety and production requirements.

Through the years, new Fieldbus Ethernet based on protocols were developed. Ethernet connection on industrial devices allows bidirectional connections from different systems to the control device, creating endless new possibilities of data acquisition and control strategies.

The introduction of classical IT and network technologies in industrial scenarios lead to a transformation in maintaining the traditional industry requirements, as previously is mentioned, and adding new elements to the IT and network technologies. The concept of network automation for industry digitalization is tailored by such transformation and the convergence between IT-OT, which creates a new paradigm with risks and opportunities for network automation.

4.4.1 IT: OT Convergence in Automation Networks

During The Third Industrial Revolution, the automation introduced into the factories through PLCs and electronics can control the entire process and allow the companies to save time and money. Those advances established the OT level, which includes levels 0, 1, and 2 from the Automation Pyramid explained in the previous section.

Automation development focused on factories, creating their development environment. Some companies centralized their resources to improve the software and hardware required for factory operation.

Some programming languages, hardware, and software were specially created and developed for factory automation. Nowadays, this development continues and allows the final users, supervisors, and operators to control the factories in a way that was not possible 30 years ago.

It is defined as Information Technologies (IT), the hardware and software for business applications. Its development has been considerably higher than OT during the same period.

4.4.1.1 Architecture Evolution

Companies' main goal is to control all the information to be able to make the best decisions in all the possible areas involved, like marketing, Human Resources, or production. Until now, the factories were connected to the enterprise control layer but not integrated. The information gathered from the control system did not allow to perform real-time decision on the plant operation.

In the Figure 4.5 is defined as the typical architecture of factories where there is no direct connection between the OT and IT. The PLC is creating a physical separation between the field components and the servers (where usually is defined that the IT level starts).

> The use of the Manufacturing Execution System (MES) or the Enterprise Resource Planning (ERP) layer has allowed this communication between the highest level of the enterprise (management) and the lowest layer of the factory (operation). One server has been used to allow bidirectional communication from PLC and MES. This communication was limited by time and space. It implies a high cost in hardware for extra storage and high-speed communications connections. Besides, the field components were limited due to sensor capabilities and communication bandwidth.

As it was indicated in the section "Enabling technologies for the new productive model", this situation is changing. The new technologies allow the connection from the operational technology level to the information technology level. New smart

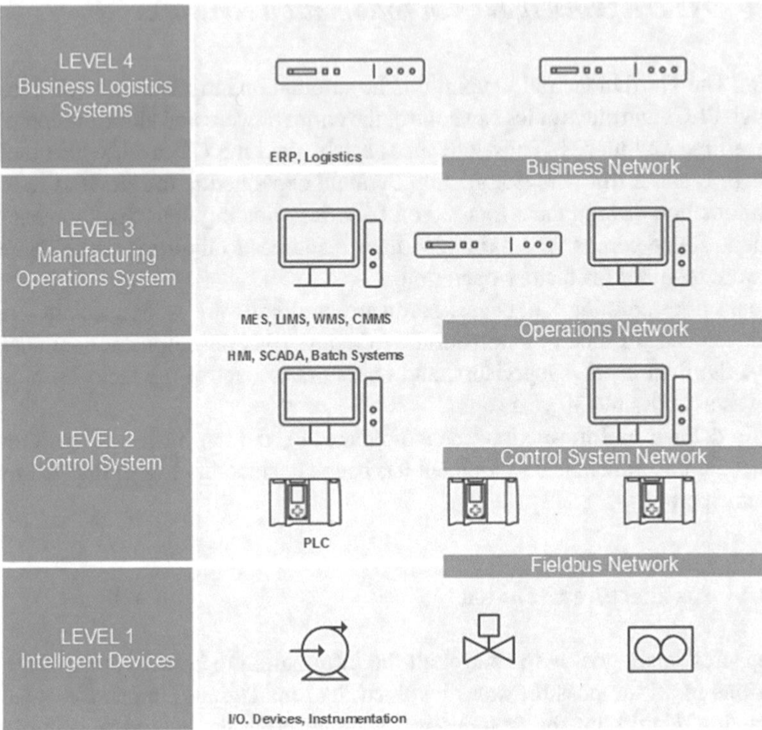

Fig. 4.5 Control system levels

field devices can provide more information about its environment, and the latest communication networks also allow them to share that information with the PLC or even through the IIoT, with the upper layer. The information can be stored in the cloud without storage limitation, which can be processed with the new technologies, for instance, Big Data, Artificial Intelligence, to find out some situations that were hidden until now.

New architecture showed in Fig. 4.6 represents the convergence between IT and OT. The connection from components to the IT level directly allows everybody to have access to more information.

The pyramid maintains functionality, but there are different changes in the different levels:

- *Level 1:* Intelligent Devices. The number of devices increases exponentially.
- *Level 2: Control level.*
  - The control devices distribution increases. There are control strategies distributed on the I/O modules instead of on the PLC. It is possible now, to perform control on L1.
  - PLCs could be software PLCs, called Virtual PLCs.

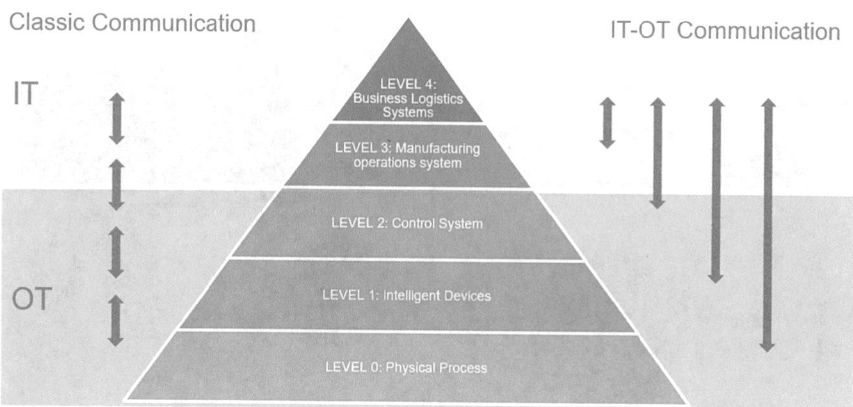

Fig. 4.6 Changes of the level architecture

- *Level 2: The monitoring level*
  - In addition to the control rooms, the monitoring function is accessible in any place thought different technologies (WEB, adapted mobile APPs. . .) and in any device (regular office laptops, tablets, smartphones. . .).
  - Secure access to the system from a regular internet connection using a VPN.
- *Level 3:* Manufacturing Operations System, the Manufacturing Execution System trend is to be integrated on the Monitoring System, to simplify the use and improve the user experience.
- *Level 4:* ERP is increasing the level of integration with the control system, requesting additional data, and sending orders to be executed by the control system.

In PERA model definition, no level above 4 were considered. IT-OT convergence increases the enterprise network and cloud relevance. This integration requires, during a control system design phase, to consider how the system will be integrated. International companies, for example, with multiple production sites, seek to connect different places to the company network. This will allow site benchmarking against other production sites.

4.4.2 *IT-OT Distribution*

Factories paradigm has changed. As we show in the pictures above, the convergence between IT and OT is real, and the boundaries are not as limited as they used to be. Companies could take advantage of new possibilities. Figure 4.7 intends to reflect the latest enabling technologies through/within the conventional OT-IT model.

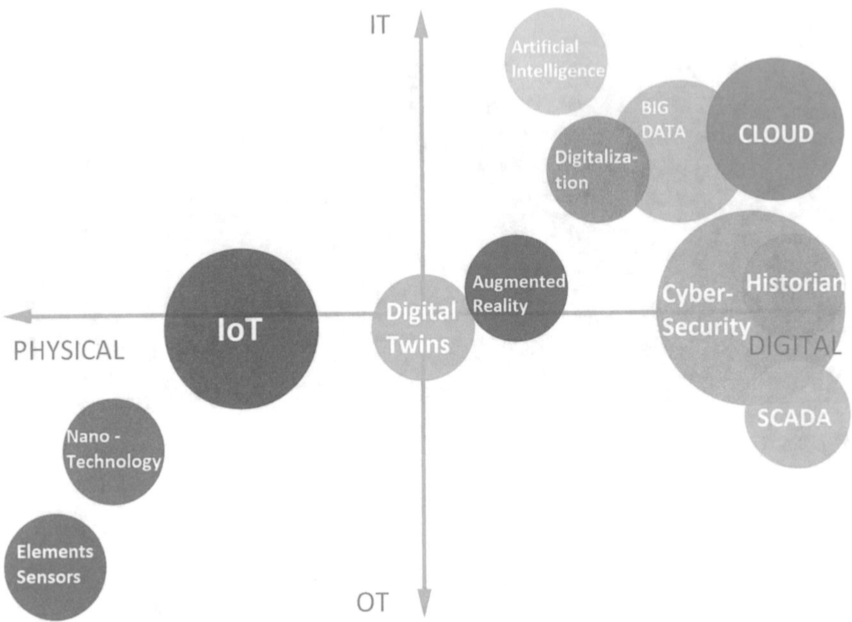

Fig. 4.7 New technologies and architectures in ISA 95

Figure 4.7 categorizes the new technologies and applications between IT-OT and Physical-Digital. This approach is made according to the current experience, but it is not fixed. The evolution of technologies could change the needs and expectations of enterprises.

Connecting IT and OT merge benefits and risks, as we will explain in other sections. We have to assess the risk but also focus on the benefits that this connection is giving to us.

Historically, IT has been excellent in data management capabilities, which allows enterprises to make better decisions regarding marketing or investments. Those decisions to maximize profitability requires a long time. On the other hand, the OT is based on real-time information. The manufacturing, scheduling, and maintenance decisions should be promptly made. When both sides are merged, the enterprise can use the factory information to make decisions that will affect all the different areas.

Since companies have access to all the factory information, they can make better decisions by analyzing in detail the development of their activities. It allows us to have better inventory control, which will reduce risk and action time.

It is important to remark the difference between having a large amount of data and the capability to extract useful information. As it was named before, new technologies, like Big Data and cloud, allow to storage many data. Then other applications as Artificial Intelligence can be used to get useful information that could be used during the decision making.

For enterprises with many factories, the benefits of this reorganization are higher. They will be able to coordinate the production of different sites all around the world once they know how each machine is working, its efficiency, its energy consumption, or its material cost. The massive amount of information all around the world will allow the enterprise to make real-time decisions that will impact different factories at the same time.

4.4.3 Reference Architecture for IIoT and Factories Digitalization

Industrial digitalization, together with the enabling technologies in the industry framework, brings new value to chain processes generating new business models and information exchange across the companies. To lay the foundations for Industry 4.0, architectural reference models are developed to standardize, classify, and describe the objects that are framed in the Smart Factories.

Reference architectures models have been established to converge the physical and digital world, focusing on the convergence of Information Technology (IT) and Operational Technology (OT), previously described [18].

The two most relevant reference architecture models are Reference Architecture Model Industrie 4.0 (RAMI4.0), developed by Industrie 4.0 workgroups, and the Industrial Internet Reference Architecture (IIRA) result of the work of the Industrial Internet Consortium (IIC).

RAMI4.0 and IIRA, while sharing the same goal of physical and digital convergence, approach different scopes and objectives. The main difference in scope, as it is shown further, is that RAMI4.0 focuses on manufacturing factories incorporating the life cycle and hierarchy levels. On the other hand, IIRA approach IIoT in overall industries such as energy, medical care, smart grid, or smart transportation regardless, at the present time, of the life cycle and hierarchy levels. Reference architecture models are under development, and both IIC and Industrie 4.0 collaborate to map and align ideas to increase the value of Smart Industries and IIoT Systems, as shown in Fig. 4.8 [15].

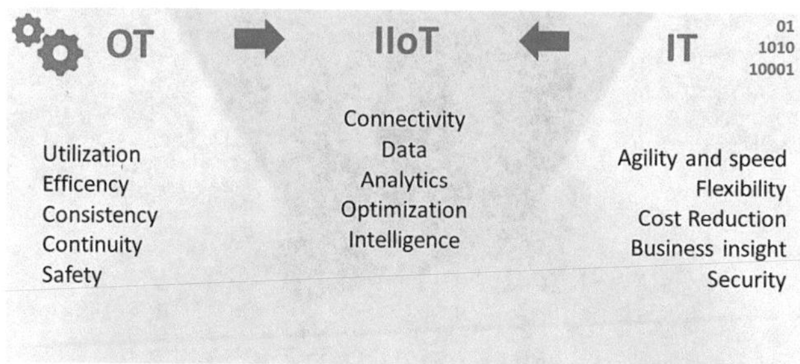

Fig. 4.8 IIoT as a transformational force driving the convergence OT and IT [15]

The Reference Architecture Model Industrie 4.0 is described below, because it is the most faithful to the industrial reality of the food and beverage factories.

4.4.3.1 Introduction to Reference Architecture Model Industrie 4.0

The Reference Architecture Model Industrie 4.0 is a three-dimensional model where differentiation between the physical world and the digital world of objects in the Smart Industry is proposed. They are divided into layers and classified according to the stage of the life cycle and their hierarchical levels as shown in Fig. 4.9.

The vertical axis is the differentiation between the real and digital worlds in six layers. At the bottom are all the physical objects that are integrated into the information systems through digitalization. The life cycle and value stream are located on the second axis. Each object is shown in its lifetime from its development to its disposal. Finally, in the third axis, the hierarchy levels are shown, in which two differentiating items of Industrie 4.0 have been added to the traditional pyramid, described above: the product and the connected world. In this way, an object can be classified and analyzed according to these three axes to place it correctly in a cube within the Reference Architecture Model Industrie 4.0 .

The most relevant criteria proposed for Industrie 4.0 products are:

- The unique identification of each physical device and its data.
- The cross-sectional security to all layers.
- The communication that the new services can use.
- The standardization of functions that make each object is the same or independent.

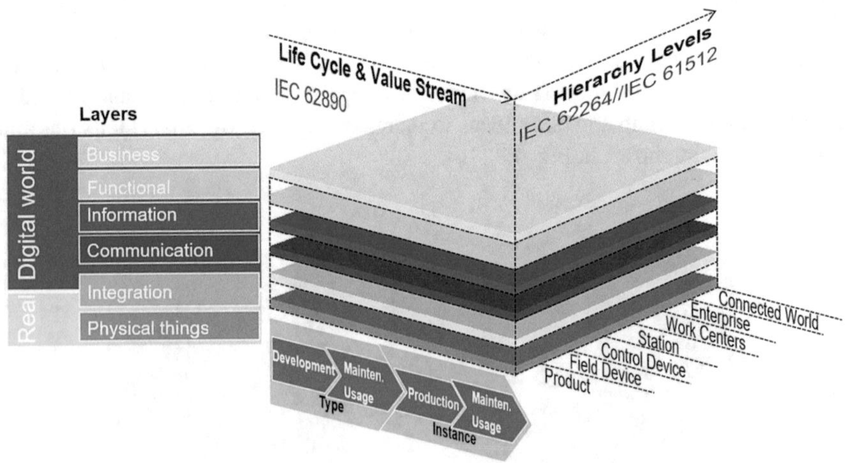

Fig. 4.9 Reference architecture model for RAMI4.0 [23]

- The proposal of a common language in which the syntax and vocabulary is univocal and allows all the agents that are related to Industrie 4.0 to know exactly the characteristics and functions of each object.

Layers

Each layer contains properties of certain assets or groups of assets. Interactions can be between layers or within the layers themselves, but a layer can't be omitted. A brief description of each layer is shown below:

- *Business layer:* Represents the commercial perspective. On this layer, business models, finance conditions, general organizational items, or connections between different business processes are located. It also represents the orchestrating of the functional layer.
- *Functional layer:* Contains the technical functionalities of the asset, including functional descriptions and runtime environment of applications and services, giving support to business processes.
- *Information layer:* Defines the data that is used and modified from the asset. This layer captures events generated and enabled for the functional layer. Besides, ensure the integrity of the data and contains the formal description of the models and rules.
- *Communication layer:* Fig. 4.10 shows the seven sub-layers of which it is composed. The physical, data link, network, transport, session, and presentation layers, where communication standards are provided. The information layer uses data and events. Commands to the integration layer are supplied.
- *Integration layer* is the Digitization of all physical assets to the information world. It generates the events of the assets and their ability to communicate. Human-machine interfaces, sensors, IT-Devices, control systems, QR-code readers are located on this layer.

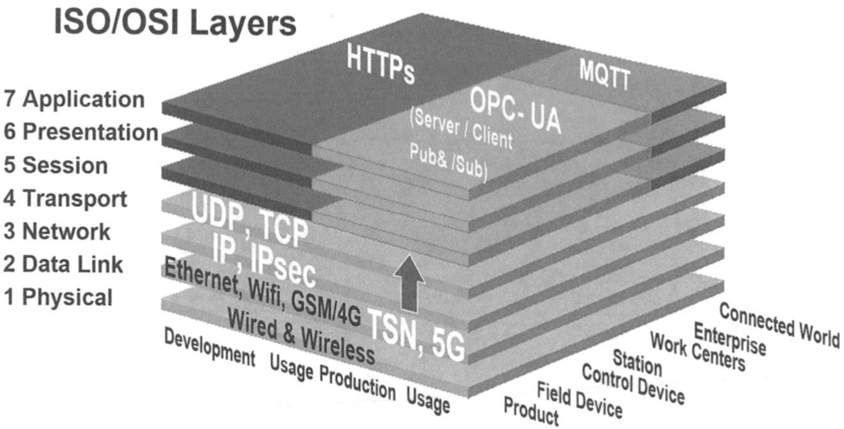

Fig. 4.10 RAMI 4.0 communication layers mapped to the ISO/OSI Model [23]

- *Asset layer:* Represents reality, the physical world. It includes devices such as sensors, actuators, components, products, cables, project plans, documentation, software, tools, and human resources, among others.

Life Cycle
The life cycle axis is divided into types and instances of each type. The type space is classified by development and use or maintenance periods and each instance by production and use or maintenance. This axis encompasses the first idea and design to decommissioning.

Hierarchy Levels
The hierarchy levels axis has been developed according to DIN EN62264-1 (IEC 62264-1) and DIN EN 61512-1 (IEC 61512-1) standards, as shown in Fig. 4.11.

As stated previously, the additional items for Industrie 4.0 are:

- *Connected world:* Represents the network between factories in which there are communication and a relationship between assets and partners.
- *Enterprise:* As an organization of commercial production and sale of goods and services. They create products to sell to the governments, the public or other entities. Enterprise Resource Planning (ERP) systems belongs to this level.
- *Work Centers:* Production center where workers, machines and operations can run simultaneously according to the products. Manufacturing Execution Systems (MES) is highlighted in this level.
- *Station:* Production unit with a main operation. They are subsections of the factories, for instance drying station.
- *Control Devices:* Pieces of equipment which mission is to control the operations. PLCs, human machine interfaces and control software are located in this level.

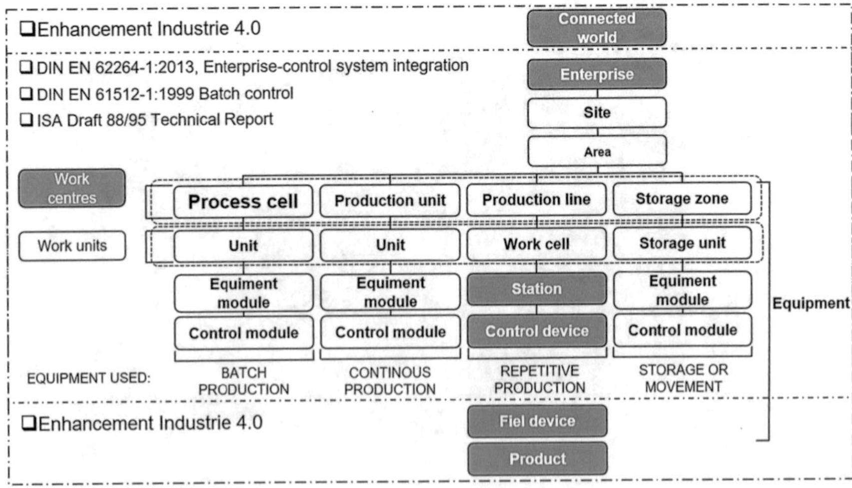

Fig. 4.11 Hierarchical levels of RAMI4.0 [23]

- *Field Devices:* Pieces of equipment in the field of measuring and acting over an operation. Sensors and actuators as valves or pumps are typically located in this level.
- *Product:* As a highlighted part of the manufacturing process, which adds value to the industry.

4.4.4 Cybersecurity

John Chambers, a former CEO of Cisco, once commented, "There are two types of companies: those that have been hacked, and those who don't yet know they have been hacked." And it seems that he was not going in the wrong way. Kaspersky, the cybersecurity and one of the most popular anti-virus provider company, this year published in his annual report, that the 42% of the industrial equipment, classified as Industrial and Control System (IACS), which use their software blocked malware attacks [11].

This, coupled with a growing industrial and critical infrastructures digital transformation, has led to both international standardization organizations such as International Organization for Standardization (ISO), International Electrotechnical Commission (IEC), International Society of Automation (ISA), North American Electric Reliability (NERC) and public institutions such as National Institute of Standards and Technology (NIST), have published a set of good practices guides, and standards (NERC CIP, ISA99, IEC 62443, NIST 800, NIST Framework) that compliment to each other's in order to create an awareness and to facilitate the security measures adoption to reduce the risks and their impacts of possible attacks to improve the organizations, infrastructures and government resilience.

Although so far, there was no specific regulation about it, governments and organizations are accelerating the implementations of new regulations like the General Data Protection Regulation (GDPR), to ensure the implementation of security measures that protect them against possible cyberattacks. Another important factor that is making the organizations to take into consideration the cybersecurity is the fact that some insurers are increasing their insurance fees due to these new risks, which makes profitable the investment in Cybersecurity not only as a preventive measure but also justifying the direct Return of Investment (ROI).

Each company, organization, or infrastructure has its own specific needs in terms of Cybersecurity, since its business, financial, industrial, Health and Safety, physical, and environmental risks are different. That is why there is no specific direct procedure or regulation to fit or apply to cover those risks. Organizations should design their own Cyber Security Management System (CSMS) or strategy to ensure their business continuity or operation, with the help of guides and standards as the Framework for improving Critical infrastructures Cybersecurity from NIST or the ISA99/IEC62443 [5].

According to the NIST definition, their Cybersecurity Framework (CSF) is an optional guide based on standards, guidelines, and exiting procedures to support

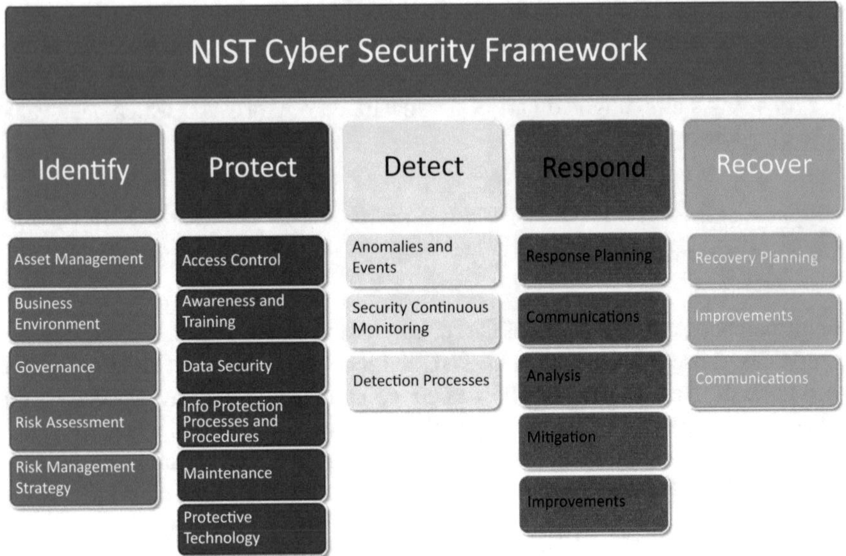

Fig. 4.12 NIST CSF core functions [20]

critical infrastructures and organizations to manage and reduce the Cybersecurity risk. It was also designed to encourage and standardize risk management communications between internal and external stakeholders [20].

As Fig. 4.12 shows, the NIST framework (CSF) core is divided into five main functions (Identify, Protect, Detect, Respond, Recover) that must be executed concurrently and continuously, as defined in its company Cybersecurity strategy (CSMS), understood within the framework of continuous improvement:

- *Identify:* Create a company Cybersecurity strategy in line with its business needs, based on its logical and physical asset inventory as staff, systems, data, devices, and capacities contextualized within its internal and external business environment, to be able to make a risk assessment and establish the strategy the face them.
- *Protect:* It allows to develop and implement the countermeasures needed to limit or reduce the impact of a potential Cybersecurity event.
- *Detect:* It allows to develop and implement suitable activities to identify the occurrence of a Cybersecurity event through permanent monitoring.
- *Respond:* Allows the definition and deployment of activities to react to an identified Cybersecurity event and mitigate its impact.
- *Recover:* Allows the deployment of activities to manage the resilience and the normal operation recovery after an incident.

In order to define the strategy, it is recommended to assign a Cybersecurity committee that will be responsible for the strategy development, its correct fulfillment, and its procedures monitoring through all the company levels and scopes. Therefore,

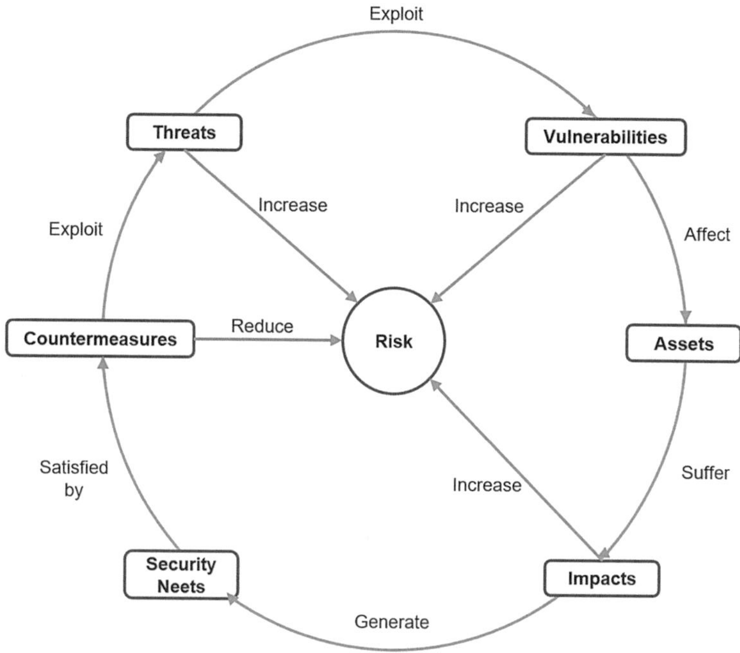

Fig. 4.13 CCI: risk analysis [5]

it is essential to well delimit its CSMS scope in terms of components and procedures to avoid misunderstandings and even being an impediment to the business. It's equally important to define the relationship between the different parts within the scope and the auxiliary elements that, despite not being within the scope, must be considered in the system due to its dependence on them.

The fundamental elements to be analyzed and documented in every risk assessment are as shown in Fig. 4.13:

- *Assets:* System resources o data needed for the correct organization performance.
- *Threats:* Events that may trigger an incident within the organization, causing material damage or immaterial losses in its assists. These could be accidental or deliberate.
- *Impact:* Consequence over an asset, that may cause a certain threat.
- *Vulnerability:* Estimation of the effective exposure of an asset to a threat, determined by the frequency of occurrence and degradation caused.

Industrial environments are extensive, complex, and diverse. Documents such as IEC 62443 recommend the use of the concepts of zones and conduits to be able to create groupings of logical or physical devices that have the same security requirements and simplify the different security functions that apply to them while defining the communication mechanisms between the different areas. In the risk assessment, once the elements described above have been identified, the tolerance

TIER	Risk Management Process	Integrated Risk Management Program	External Participation
Partial	• Not Formalized • Reactive	• Limited awareness • Irregular risk management • Private information	No external collaboration
Risk Informed	• Approved practives • Not widely use as policy	• More Awareness • Risk-informed, processes & procedures • Adequate resources • Internal Sharing	Not formalized to interact & share information
Repeatable	• Approved as Policy • Updated regularly	• Organization approach • Risk-informed, processes & procedures defined & implemented as intended, and reviewed • Knowledge & skills	• Collaborate • Receive information
Adaptive	• Continuous improvement	• Risk-informed, processes & procedures for potential events • Continuous awareness • Actively	Actively shares information

Fig. 4.14 NIST implementation TIER [20]

to these risks must be calculated and the degree of these (Risk = Criticality x Likelihood x Impact) to prioritize and design the most effective protection measures [3].

In industrial environments, the complete risk elimination is virtually impossible, so the necessary countermeasures should be sought to minimize the impact of possible threats to a tolerable level. It is important to detect the level of risk tolerance and when a countermeasure reports benefit or is contraindicated due to cost or availability. Where appropriate, for those unavoidable risks or whose countermeasures are not viable, companies may choose to take out insurance that covers them [5].

When measuring how close the Cybersecurity risk management system approaches to what is defined in the CSF, four levels or tiers are defined based on the elements of Fig. 4.14:
Risk Management process, Integrated Management Program and External Participation. With this, when establishing the roadmap or objectives of the Cybersecurity strategy, it must be analyzed the current profile and set a target profile as well as the action plan to achieve it.

4.5 Motivation for Industrial Network Transformation

This chapter examined the path to the fourth industrial revolution. The industrial sector seeks to reach a new productive model. The motivations of the Fourth

Industrial Revolution are cost reduction, rapid technology adoption, end-product customization, and product delivery time reduction.

These requirements are not new, but new enabling technologies facilitate now to implement solutions that will meet those requirements with time reduced and cost-effective implementation. The barriers to develop new businesses are lower than before.

Industrial transformation begins with the Industrial reference architecture evolution. This architecture did not suffer remarkable changes in the last four decades. Today there is a unique situation with significant opportunities to seize and risks to mitigate.

The reference architecture's most significant modification is the IT-OT convergence. OT systems incorporate IT new procedures, and new working methodologies, like Cybersecurity. IT engineers are now involved in OT design, to balance the system requirements and facilitate the overall integration. Maybe, one day, the sensor will be connected directly to the enterprise network and will be considered an IT asset.

These changes will enhance horizontal and vertical communications. Peer to Peer, Machine to machine (M2M) communication will be strengthened. Communication cadence between different business levels (ERP to Industrial Devices) will evolve from transactional to near real-time information, which will launch data-driven decisions to the next level.

Peer to peer communications will increase. The hierarchy will be softer.

System integrators are now forced to expand their traditional scope to the top levels of the automation pyramid. IT software providers are focused on the lower pyramid levels, encouraged by the IIoT new possibilities and market requests to deliver vertical solutions. This situation creates competition, and companies in the industrial sector see their position challenged by regular software companies.

Machine builders will fight to avoid to become a commodity. During the first three industrial revolutions, industrial production machines were the core of the facilities. Now, during the Fourth Industrial Revolution, the data could compete to be the industrial core and could displace machinery.

> The future looks promising. Industry 4.0 and IIoT initiatives were born to improve twenty-century factories, to improve their productivity through the new enabling technologies. This competition creates a gameboard where IT companies, OT system integrators, machine, and sensor manufacturers are looking for their role. Of course, no one is willing to be out of the game.

References

1. Alcober, V.: Enrico Fermi y los primeros reactores nucleares americanos. Sociedad Nuclear Española (SNE), Madrid. ISBN 9788469709665 (2014)
2. ANSI/ISA-95.00.01-2010 (IEC 62264-1 Mod) Enterprise-Control System Integration - Part 1: Models and Terminology
3. ANSI/ISA 99 - IEC 62443: Security for Industrial Automation and Control Systems, several parts: 1-1(2017), 2-1(2009), 3-2(2017), and 3-3 (2013)
4. Brynjolfsson, E.: Big data: the management revolution. A. McAfee. Harvard Bus. Rev. **90**(10), 60–68 (2012)
5. CCI: Guía para la construcción de un SGCI, Sistema de Gestion de la Ciberseguridad Industrial (2014)
6. Cleland, S.: Google's "Infringenovation" Secrets https://www.forbes.com/sites/scottcleland/2011/10/03/googles-infringenovation-secrets/393e01c230a6
7. Deane, P.: The First Industrial Revolution, 2nd edn. Cambrige University Press, Cambrige (1979). ISBN 0521226678
8. Evans, D.: The Internet of Things How the Next Evolution of the Internet Is Changing Everything. White Paper Cisco Internet Business Solutions Group (IBSG) (2011)
9. Industrial Internet Consortium: The Industrial Internet of Things.Volume G1: Reference Architecture. Version 1.9. (2019)
10. ITU-R WP5D: IMT vision– framework and overall objectives of the future development of IMT for 2020 and beyond. Rec. ITU-R M.2083-0 (2015)
11. Kaspersky Industrial CyberSecurity: solution overview (2019)
12. Laughton, M.A., Warne, D.F.: Electrical Engineer's Reference Book, 16th edn. Butterworths, London (2002). ISBN 9780080523545
13. Lee, J., Kao, H.-A., Yang, S.: Service innovation and smart analytics for industry 4.0 and big data environment. Proc. CIRP **16**, 3–8 (2014)
14. Levin, M.R., Forgan, S., Hessler, M., Kargon, R.H., Low, M.: Urban Modernity Cultural Innovation in the Second Industrial Revolution. The MIT Press Cambridge, Massachusetts (2010). ISBN 9780262013987
15. Lin, S.-W., Murphy, B., Clauer, E., Loewen, U., Neubert, R., Bachmann, G., Pai, M., Hankel, M.: Architecture alignment and interoperability. An Industrial Internet Consortium and Plattform Industrie 4.0 Joint Whitepaper (2017)
16. Marsch, P., Bulakci, Ö., Queseth, O., Boldi, M.: 5G System Design: Architectural and Functional Considerations and Long Term Research. Wiley, New York (2018). https://doi.org/10.1002/9781119425144
17. Microsoft Azure: Introduction to cloud computing. https://https://azure.microsoft.com/en-us/overview/
18. Minoli, D.: Enterprise Architecture A to Z. Frameworks, Business Process Modeling, SOA, and Infrastructure Technology. CRC Press, Boca Raton (2008)
19. Mokyr, J.: The Bristish Industrial revolution. An Economic Perspective, 2nd edn. Westview Press, Boulder (1999). ISBN 081333389X
20. National Institute of Standards and Technology (NIST): Framework for Improving Critical Infrastructure Cybersecurity v1.1 (2018)
21. NGMN: White Paper, 5G White Paper (2015)
22. Onar, S.C., Ustundag, A.: Smart and Connected Product Business Models. Springer, Berlin (2018)
23. Adolphs, P., Berlik, S., Dorst, W., Friedrich, J., Gericke, C., Hankel, M., Heidel, R., Hoffmeister, M., Mosch, C., Pichler, R., Rauschecker, U., Schulz, T., Schweichhart, K., Steffens, E.J., Taube, M., Weber, I., Wollschlaeger, M., Mätzler, S.: Reference Architecture Model Industrie 4.0 (RAMI4.0). Plattform Industrie 4.0. English translation of DINSPEC 91345:2016-4 (April 2016)

24. Rifkin, J.: The Third Industrial Revolution How Lateral Power Is Transforming Energy, the Economy, and the World. Palgrave Macmillan, New York (2011)
25. Taylor, G. R.: The Transportation Revolution, 1815–1860., vol. 4. Holt, Rinehart and Winston, New York (1962). ISBN 0873321014
26. Telegram from Orville Wright in Kitty Hawk, N.C., to his father announcing four successful flights, 1903 Dec. 17 Library of Congress Washington, D.C. 20540 USA. Reproduction number LC-USZ62-65459. http://hdl.loc.gov/loc.pnp/cph.3b12982
27. The Nobel Prize in Physics 1956. NobelPrize.org. Nobel Media AB 2019. 31 Oct 2019. https://www.nobelprize.org/prizes/physics/1956/summary
28. Willems, L.: On the Supply Chain in the Fourth Industrial Revolution. Université Catholique de Louvain, Louvain-la-Neuve (2018)
29. Williams, T.J.: The Purdue Enterprise Reference Architecture: A Technical Guide for CIM Planning and Implementation. Instrument Society of America, Research Triangle Park (1992)
30. Witkowski, K.: Internet of Things, Big Data, Industry 4.0 – Innovative Solutions in Logistics and Supply Chains Management. Procedia Engineering, vol. 182, University of Zielona Gora, Poland (2017), pp. 763–769

Part III
Cyber Security

Chapter 5
Security in Decentralised Computing, IoT and Industrial IoT

Monjur Ahmed, Sapna Jaidka, and Nurul I. Sarkar

5.1 Introduction

Computing approaches are going through a phase shift with new computing and technologies are being introduced. New computing paradigms, for example, Cloud Computing, IoT, IIoT are getting momentum. The implementation of these computing approaches through distributed and decentralised architectures are taking the computing and its meaning into a completely new definition and dimension. The architectural and implementation strategies and factors computing infrastructures are getting complex and challenging due to a major shift in computing.

A trend to include all kinds of devices with computing power to a greater Internet-based network is underway. At the same time, leveraging the use of remote computing infrastructure to reduce complexity and total cost of ownership for end-users' is also evident. Towards the path of such endeavour, Cloud Computing and IoT gained momentum. Cloud Computing denotes using remote and shared computing resources or infrastructure [7]; and IoT refers to an Internet-based network of gadgets (e.g. objects embedded with electronics, software, sensors) [10, 28] with computing capability that people use in their daily life. Such gadgets range from information acquiring sensors to household devices (e.g. phones, microwave, washing machine and so on). Variants of these computing approaches have also emerged. For example, Cloud Computing is used with IoT [21] and termed as Cloud-based IoT. Another example is IIoT, which is a variant of IoT but considers IoT deployment within industrial context. Regardless of being identified as a variant

M. Ahmed (✉) · S. Jaidka
Waikato Institute of Technology, Hamilton, New Zealand
e-mail: monjur.ahmed@wintec.ac.nz; sapna.jaidka@wintec.ac.nz

N. I. Sarkar
Auckland University of Technology, Auckland, New Zealand
e-mail: nurul.sarkar@aut.ac.nz

I. Butun (ed.), *Industrial IoT*, https://doi.org/10.1007/978-3-030-42500-5_5

or not, the core aspects, features, and challenges remain same. IIoT comes with its potential for many industries [15].

While the above computing approaches bring flexibility to the end-users, they bring complexity at the infrastructure level. With all challenges and pros & cons, the security concerns (i.e. vulnerabilities, threats) of the above computing technologies are also becoming more apparent. Security is one of the main concerns in Cloud Computing and IoT. These computing means tend to be distributed at the infrastructure level. From security perspective, the distributed resources in Cloud Computing and IoT makes it more complex with added concerns. As a result, security for Cloud Computing and IoT (along with all their variants) require security to deal with distinct concentration and importance.

Considering IIoT security, the security concerns of Decentralised Computing, Cloud Computing as well as IoT enter into the picture. This is since IIoT is essentially a flavour of IoT, Decentralised Computing is a feasible approach to design and implement IoT infrastructure, and Cloud Computing is a major contributing factor in contemporary computing infrastructures including IoT and IIoT. This chapter aims to introduce the concept of Decentralised Computing, Cloud Computing, and IoT; and eventually the security aspects of Cloud Computing, IoT and IIoT.

IIoT is the industrial context-specific version of IoT. IIoT thus may embed into real time processes and systems which may not be necessarily be the case for conventional IoT. As a result, integrity and security concerns for IIoT when compared to those in IoT, are more connected to safety-critical and mission critical factors.

In this chapter, we first look into the concepts of Cloud Computing, Decentralised Computing, IoT and IIoT. The following discussion addresses security challenges in Decentralised Computing, followed by security in Cloud Computing and in IoT. Different factors related to IIoT security issues and challenges are also investigated.

5.2 Cloud Computing, Decentralised Computing, IoT, IIoT

Cloud Computing is embedded into almost every aspect of computing. IoT and IIoT architectures are also incorporating Cloud Computing. As a result, Cloud Computing and Decentralised Computing warrants discussion in IoT and IIoT security context. In Cloud Computing, resources are distributed. From security perspective, a distributed architecture may achieve better security if decentralised. Being a variant of IoT and inheriting all aspects of IoT, an IIoT architecture may also incorporate features of decentralised and distributed computing, and Cloud

Computing. In this section, we discuss the concepts of Decentralised Computing, Cloud Computing, IoT, and IIoT.

5.2.1 Decentralised Computing

Decentralised Computing refers to a computing setting or architecture which has no core or central controlling entity. A Decentralised Computing setting is ideally made up of nodes or computers which collectively define the core of the architecture, but any given node alone bears no significant controlling or administering power over the architecture.

In a distributed architecture, soft and hard resources are distributed. In decentralised architecture, the resources are also essentially distributed. What makes Decentralised Computing distinct from distributed computing is that, a decentralised architecture is necessarily a distributed architecture but not vice versa.

Figure 5.1 illustrates the concept of distributed and decentralised architecture assuming an application made up of three components (or parts)—a circle, a rectangle and a triangle. Referring to Fig. 5.1, a distributed architecture may have a system (e.g. an application) distributed in different networks and computers by creating multiple copies of itself. But in decentralised architecture, a requirement is to scatter the elements of a system into different computer and networks in such a manner that there is no single core of the system. We agree with Ahmed [1] and argue that a purely decentralised system should also distribute the elements of a system in such a manner so that no single entity (e.g. a server, an application container or a node) of the network may have all the elements of a system, as

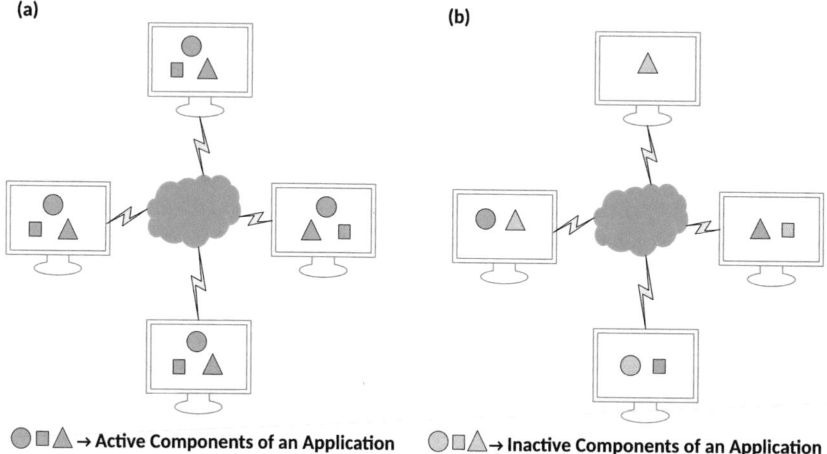

Fig. 5.1 (**a**) Distributed and (**b**) Decentralised architecture

illustrated in Fig. 5.1. This is to ensure context illiteracy for an attacker through the avoidance of the provision to build a holistic picture of the system. In a distributed architecture, all distributed components may be active. In a decentralised architecture, some components may be inactive and may become active as and when required—example of such approach can be found in [1].

A decentralised architecture eliminates a single core of a system. A purely decentralised architecture has no central controlling part of a system. This eliminates a single point of failure for a system. Besides, the decentralised elements may have redundant copies that are scattered across the whole architecture. Such redundancy helps a decentralised architecture to maximise service availability compared to its centralised counterpart.

5.2.2 Cloud Computing

Cloud Computing warrants discussion within IIoT context in that, Cloud Computing is a computing approach being embraced by all computing approaches and technologies [16], including IIoT. Cloud Computing or simply "the Cloud", is a means to host and deliver on-demand computing resources or services via a public infrastructure (e.g. the Internet). The services vary from elementary services like computation and data storage to more advanced services like machine learning. There is no doubt that Cloud Computing is making a significant impact on Information and Communications Technology (ICT) and is one of the most significant computing approaches to date. Scalability, deployment, automation, affordability, and manageability are the assets that Cloud offers to businesses as well as individuals. It has become a critical part of IT infrastructure for both large scale and small-scale businesses. Cloud Computing comes with its own pros and cons. For an IIoT architecture, using remote services or application through public infrastructure (i.e. the Internet) makes an IIoT a Cloud-based IIoT. As a result, security concerns in Cloud Computing may be inherited into an IIoT architecture.

From an operational perspective, Cloud Computing and IIoT may overlap to certain extent. Cloud has different meaning to different parties depending on their experience and background. Different definitions of Cloud Computing exist. According to Buyya et al. in [12], a Cloud is a type of parallel and distributed system consisting of a collection of interconnected and virtualised computers that are dynamically provisioned and presented as one or more unified computing resources based on service-level agreements established through negotiation between the service provider and consumers. Wang et al. in [29] defines Cloud Computing as a set of network-enabled services, providing scalable, Quality of Service (QoS) guaranteed, normally personalised, inexpensive computing infrastructure on demand, which could be accessed in a simple and pervasive way. The above perception of Cloud Computing makes it apparent that Cloud Computing and IIoT architectures may essentially have overlapping features.

Despite many proposed definitions, the definition given by the National Institute of Standards and Technology (NIST) is widely accepted. They define Cloud Computing as "a model for enabling ubiquitous, convenient, on-demand network access to a shared pool of configurable computing resources (e.g., networks, servers, storage, applications, and services) that can be rapidly supplied and released with minimal management effort or services provider interaction" [22].

5.2.2.1 Cloud Models

Cloud Computing is perceived from two distinct sets of models: Service oriented models with three sub-models, and deployment–oriented models with four sub-models.

1. **Service model:** These models refer to the type of services that can be accessed on a Cloud Computing platform. The three service models are:
   - **Infrastructure as a Service (IaaS):** IaaS is the elementary infrastructure that provides virtual resources like virtual storage, virtual machines, hardware and networks. Users are responsible for assembling their virtual resources. The main advantage of IaaS is that the client companies avoid the costs of purchasing and maintaining their own hardware and software.
   - **Platform as a Service (PaaS):** This model has a limited scope as compared with IaaS. In PaaS model, the platform is provided to a user on which a user can deploy, install or manage any application or service. A user does not have any control over the platform but can use the platform to achieve their goals.
   - **Software as a Service (SaaS):** In the SaaS model, the application or software is provided to the user by the Cloud provider, and the user's responsibility is limited to managing and accessing the respective application or software. Everything from the application down to the infrastructure is the vendor's responsibility.
2. **Deployment model:** These models refer to the location of the Cloud and the purpose of the Cloud. The four deployments models are:
   - **Public Cloud:** The infrastructure of a Public Cloud if normally provisioned by third party entity known as Cloud Service Provider (CSP).
   - **Private Cloud:** The Private Cloud model is operated specifically for a particular organisation, and is accessible and controlled by that organisation or a third party.
   - **Hybrid Cloud:** A Hybrid Cloud model is a combination of private and public Clouds.
   - **Community Cloud:** The Community Cloud model is shared by more than one organisation to serve a common purpose.

The definition and models (service and deployment) of Cloud Computing makes it apparent that an organisation using Cloud Computing may house their sensitive and confidential digital assets on CSP's IT infrastructure which is under the ownership of the respective CSP. This may lead to security and privacy concerns. Subsequently, Cloud Computing demands scrutiny and exploration for Cloud-based IoT (and thus, for Cloud-based IIoT).

5.2.3 IoT

As an initial note for this section, and as mentioned earlier, all aspects of IoT equally applies to IIoT. IIoT is a flavour of IoT inheriting all aspects of IoT in addition to its own features. The discussion in this section on IoT applies to IIoT as well, bearing the fact in mind that IIoT is IoT in industrial context.

IoT is an emerging technology [19]. IoT can be defined as a computer (any devices with computing capability, to be more precise) networking technology that encompasses distinct physical objects or 'things' with different capabilities and communication methods to exchange information with themselves, with their external environment or both. There are several meanings or definitions given to the term IoT depending on who is defining it. IoT has added new dimension to everyday life by connecting smart things [23]. According to IEEE [6] "the IoT is a system consisting of networks of sensors, actuators, and smart objects whose purpose is to interconnect "all" things, including every day and industrial objects, in such a way as to make them intelligent, programmable, and more capable of interacting with humans and each other." Examples of these devices (or objects or things) could be any device that can be connected to a network such as smart watches, sensors etc. These objects are connected to each other via internet and are able to communicate and exchange information, while operating in a vulnerable environment. This leads to major security challenges. the layered architecture of IoT come with security issues [11].

The most important part of IoT is communication. Distinct devices should communicate with each other in order to interconnect. The devices communication can be done in several ways; direct communication, communication via gateway through communication network or communication without gateway through communication network as shown in Fig. 5.2, which can also be considered as the overview of an IIoT architecture. In Fig. 5.2, 'a' shows communication of devices via a gateway through Internet, and 'b' shows that two devices are communication over Internet without a gateway, and 'c' illustrates direct communication (via Bluetooth) without Internet. It is also evident from the Fig. 5.2 that a physical thing can be mapped into the information world via one or more virtual things, while virtual things do not necessarily need to be associated with any physical thing and can exist independently of any physical existence.

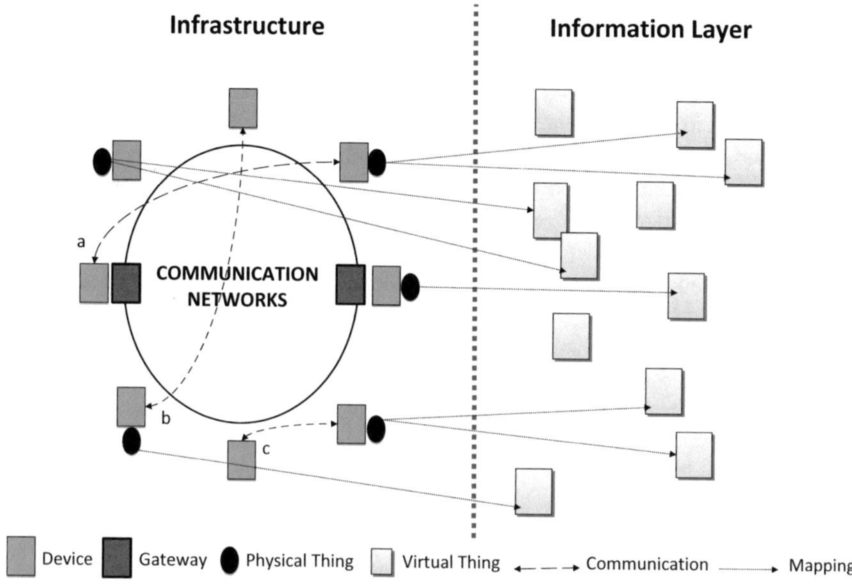

Fig. 5.2 An overview of IIoT (adapted from [26])

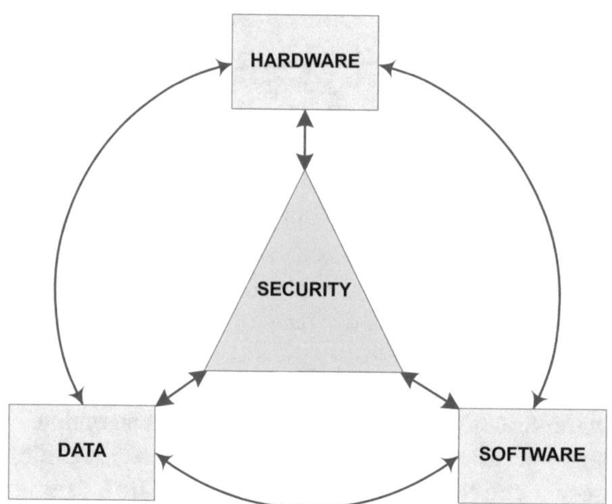

Fig. 5.3 Building blocks of IIoT

Figure 5.3 shows the building blocks of IoT. As illustrated, an IoT (and thus, IIoT) infrastructure incorporates hardware, software, data and the communication networking infrastructure within which hardware, software and data reside and operate. Like any other computing means, security aspects is also a building block for IoT and IIoT.

5.2.4 Industrial IoT (IIoT)

Industrial IoT (also known as the Industrial Internet or Industry 4.0) is the application of IoT in an industrial context and represents the merging of digital and real industry. IIoT aim to enhance industrial process [20]. While IoT mostly influences healthcare, transportation and smart homes, IIoT focuses on real-time industrial environments. IIoT incorporates machine learning, machine-to-machine communication, big data technologies and automation technologies. The aim is to intensify the efficiency of operations in industries. The emerging technologies like big data or Cloud Computing have brought in considerable opportunities for promoting industrial upgrades by increasing the efficiency and productivity and reducing complexities. A classic IIoT system comprises of sensors, micro-controllers, Cloud Computing infrastructure, intelligent system applications, security mechanisms etc.

IIoT is in its very initial state. Although it is growing very fast; there are various challenges attached to it that might affect its future growth. Challenges like integrating data is of major concern as it is not convenient to move from data to business value. Shortage of skilled staff in this field is another major challenge.

Another main issue for industries is security. The security issues are significantly important because of the rising vulnerability to attacks and data breaches. In the context of industrial internet of things, data would be considered as critical and sensitive because data would comprise of different aspects of industrial operation, including highly sensitive information about products, business strategies, and companies. The shift to more vulnerable and open networks and data sharing capability of IoT elevates challenges in industrial areas. A significant damage and loss can occur due to breach of sensitive data [30].

Despite of these obstacles and challenges, adoption of industrial internet of things is growing rapidly. While adoption to IIoT maybe an option; giving full attention to IIoT security in the case of deploying IIoT is a must, not an option.

We argue that Decentralised Computing approach may significantly minimise security concerns for any computing scenario, including IIoT. This is the motivation to consider security aspects of Decentralised Computing in next section. Like any other computing approach, and despite of our argument for Decentralised Computing to be more credible approach towards IIoT. It also come with its own security concerns and challenges.

5.3 Security in Decentralised Computing

Since Decentralised Computing deals with resources scattered over an architecture, the resources management becomes complex. If deployed properly, a Decentralised Computing architecture pose challenge to an attacker in that, there is no core of the system and thus compromising part of decentralised architecture maximises the probability of the attacker being exposed, instead of the architecture being gradually compromised on an architecture-wide scale. At the same time, the greatest challenge comes with decentralising the core through resource distribution and resources management for decentralised architectures.

The decentralised nature of a computing scenario makes the design of maintenance of such architecture complicated [14]. While a decentralised architecture may significantly minimise the chance of an architecture wide attack, it does not prevent partial breach through its design. The partial compromise of part of a Decentralised Computing architecture required vigilance and careful design of the algorithms and mechanisms that will employ the coordination and collaboration among the nodes of a decentralised architecture. It is virtually impossible to pinpoint technology specific security loopholes, threats or vulnerabilities for a decentralised architecture, since this is a conceptual computing approach.

Any distributed system that is decentralised also, come with security challenges entirely based on the technologies used and the technical soundness and implementation integrity of the underlying security mechanism. Cloud Computing and IoT are distributed by nature and may incorporate Decentralised Computing to any extent. The following sections explore the security aspects of Cloud Computing and IoT from a decentralised implementation viewpoint.

5.3.1 Decentralised Computing: Challenges and Security

As discussed earlier, Decentralised Computing architectures tend to distribute the elements in a manner to achieve no single point of failure. To do so, an element is further sub-divided and may also incorporate redundant copies of an element scattered across the architecture. A decentralised approach essentially inherits all security concerns as a computing network. In addition, the scattered elements and decentralisation of resources introduces few innovative challenges. The following is a non-exhaustive list of few design and security challenges for Decentralised Computing.

1. **Algorithmic Complexity:** A decentralised architecture comes with extensive processing due to redundancy and scattered elements. Algorithms for Decentralised Computing need to prioritise scattered resource management and integrity of processes over pace of performance. For a decentralised architecture, performance in terms of speed can be mitigated to some extent through maximising the computing power of the devices involved, but the lack

of integrity for all elements to work coherently may not be mitigated by any other factor.

2. **Residual Loopholes of Process–intensity:** A decentralised architecture is process intensive by nature. Some Decentralised Computing settings that use redundant processing and elements incorporate massive processing to carry out its operation. Intensive processes run in a decentralised architecture. This may lead to residual consequences of introducing security loopholes in an architecture.
3. **Security Mechanism:** In a decentralised setting, the process could be divided into sub-processes distributed over nodes—this helps the decentralised system to hide context from malicious parties, but at the same time, the collaboration of sub-processes to collectively serve remains challenge from security perspective. The countermeasures (or security mechanisms) to combat with the security concerns for the above is not straightforward. A security mechanism for a decentralised architecture also needs to be decentralised. If the cause is distributed and decentralised, the effect is also distributed and decentralised. In the same way, the security concerns for a decentralised security mechanism is decentralised, and the approach to countermeasure security measures in Decentralised Computing needs to be decentralised. Such a decentralised security model is proposed by one of the authors [1].
4. **Audit Trail:** The audit trail also needs to be distributed and decentralised. Maintaining integrity of the audit trail may be challenging due to the complexity of the decentralised architecture—this has potential to contain vulnerabilities and security loopholes. One of the beauties of Decentralised Computing system is its ability to hide the context of total computing environment, making it harder for the attacker to engineer such systems. Any kind of audit for a decentralised architecture must ensure that it does not reverse-engineer the decentralisation process to build centralised picture of the architecture. If any automated auditing system is used on a Decentralised Computing that connects the 'dots' of decentralisation to get a full picture of the architecture, compromising the automated auditing tool is enough to find different ways for the attackers to compromise the system.
5. **Decentralised Ownership:** Decentralised architecture tends to spread the control and processing of information across the architecture. This makes every node connected to a decentralised architecture a 'server' and thus more security loopholes may create. Having said that, we believe this is a mere design challenge for decentralised architecture and cannot be outweighed in terms of benefits compared to its centralised counterpart. Bring-Your-Own-Device (BYOD) can be considered as a Decentralised Computing context where end-users' devices may act not only as a client, but also may contribute to processing if the architecture it is connecting to. This may become of trend in tomorrow's computing, though a challenge related to privacy and control on the processes and data in such context is a concern–this applies both ways, that is, for the architecture BYOD devices connecting to, and for the devices. What kind of control a decentralised architecture will have on BYOD devices, and what level

of access and authorisation the BYOD devices will have to the connecting network is a security concern and a design challenge. The access by BYOD devices incorporates information security concerns [5, 25].

5.3.2 Cloud Computing Security

As mentioned earlier, Cloud Computing and IIoT may overlap and Cloud Computing security concerns may be inherited into and IIoT infrastructure. Cloud computing comes with numerous security concerns [27]. Security is probably the most delicate and biggest concern for Cloud architecture because the trustworthiness of Cloud architecture relies upon its security. Cloud Computing security comprises of a set of policies, controls, techniques and advancements that work together to protect Cloud-based systems, data and infrastructure.

The way Cloud security is delivered will rely upon the individual Cloud provider, but execution of Cloud security procedures ought to be a joint responsibility between the business owner and solution provider. In order to evaluate the risks, following need to be analysed:

- Resources that are to be moved to the Cloud.
- The sensitivity of the resources to risk. Risks that need to be evaluated are loss of privacy, unauthorized access by others, loss of data, and interruptions in availability.
- The risk associated with the particular Cloud type in terms of data location, privacy and CSP's approach to manage the digital assets of it's users'.
- Particular Cloud service model.
- The selected CSP's process including how data is transferred, where it is stored, and how to move data both in and out of the Cloud.

Considering IIoT, if the Perception Layer or the 'server and storage end' of and IIoT is located into a CSP's data centre, the above applies to IIoT as well. For Cloud Computing, and for any Cloud-based IIoT—in addition to the above list—it is also important for organisations to carry out organisational and technological feasibility prior to migrate to the Cloud [3]. Otherwise, security issue in the Cloud may bring disastrous consequence for an organisation. Cloud Computing comes with a range of technological and human factors related security concerns [2]—such security concerns emerging from technological or human factors that equally applies to IoT and IIoT contexts as well.

Security is one of the key challenges for Cloud engineering [17] and it is important to review existing difficulties. Some of the deployment models in Cloud architecture allow users to build customized environments. Cloud clients might have the option to pick their very own OS and software. However, such option may prompt security breach in Cloud architecture. In the Cloud, functionality of one level of the infrastructure (for example, SaaS) and security relies upon the strength of its lower level (for example, PaaS). This level dependency of Cloud service models makes Cloud security a delicate aspect to consider. Then again, a lower level may have little authority over the security parts of a higher level (for example, IaaS) which may prompt security breaches [13]. If part of the architecture of an IIoT is situated in the Cloud, the above become critical success factors for IIoT from a security point-of-view.

5.3.3 *IoT Security*

Security related concerns in IoT come with no surprise. The layered architecture of IoT incorporates different security and privacy threat for each layer [8]. IoT security is an area that needs to be scrutinized because of the occurrence of several breach incidents. Mostly a network is attacked by a common IoT device in a network. Therefore, it is very important to implement security measures in order to protect the network with which IoT devices are attached.

IoT incorporates security and privacy concerns [24]. Managing access control, authentication and authorisation are some of the major concerns for IoT [4]. There are numerous challenges that anticipate the security of IoT devices. Since proposition of interconnecting IoT devices is not very old, security has not been given much importance in the design phase of a device. Also, in light of the fact that IoT is an incipient market, numerous device manufacturers are progressively keen on getting their devices to showcase rapidly, rather than finding a way to incorporate security measures right at the beginning.

The utilization of default passwords or hard-coded passwords is one of the signification issues, which can cause security breaks. Even if users change the passwords, they are usually not strong enough to avert intrusion. IoT devices being resource constrained in one more challenge confronting IoT devices. Accordingly, numerous IoT devices don't or can't provide advanced security. For instance, temperature monitoring sensors can't deal with cutting edge encryption or other safety measures.

The utilization of default passwords or hard-coded passwords is one of the signification issues, which can cause security breaks. Even if users change the passwords, they are usually not strong enough to avert intrusion. IoT devices being resource constrained in one more challenge confronting IoT devices. Accordingly, numerous IoT devices don't or can't provide advanced security. For instance, temperature monitoring sensors can't deal with cutting edge encryption or other safety measures.

Trying to connect legacy assets which are not actually made for IoT connectivity is another security issue. Superseding the infrastructure of these assets with networking technology is a prohibitive expense. However, the chances of attack are much higher in these legacy assets because of the lack of updating and not having security against modern threats. As far as updates, large number of systems just incorporates help for a particular time period. For new and legacy resources, security can slip by if additional help isn't included as numerous IoT devices remain in the network for a long time.

Security of IoT is also tormented by an absence of standards acknowledged industries. Despite of existence of numerous IoT security frameworks, no single framework is widely accepted by the industries. Huge organisations and industries may have their very own particular standards, while certain portions, for example, industrial IoT, have restrictive, inconsistent standards from industry pioneers. Security of systems and guaranteeing inter-operability between them have become even harder with the assortment of these standards makes it hard to secure systems.

Manufacturers, service providers and end users must learn to see security as a mutual issue. Security and privacy of products and services should be prioritized by manufacturers and service providers. Also end users must make certain to play it safe by frequently changing the passwords, making use of available security software and installing patches.

Variety of systems exists for ensuring IoT security, but none of them is accepted by industry as a standard. However, embracing a framework for IoT security can help. organisation such as the IoT Security Foundation, GSM Association and other have release various such frameworks.

From smart home to a big manufacturing plant, IoT security hacks can occur in any industry. The seriousness of effect depends incredibly on the type of a system, the information gathered and additionally the amount of sensitive data it has. Attacks, for example, disabling the brakes of a connected vehicle, or infusing too much or too less medication to a patient by hacking the connected infusion pump can be perilous.

There are various ways to protect IoT systems and devices depending upon a place in the IoT ecosystem and particular IoT application. The common IoT security measures are:

- **Proactive Security Considerations:** IoT designers ought to incorporate security toward the beginning of any device development.
- **Avoiding Hard-coded Credentials:** A developer should make sure that the hard-coded credentials must be changed before the device starts to function. Sometimes, product has default credentials which should be replaced by strong password or some other measures like figure print detection etc.
- **API Security:** Securing APIs is fundamental to ensure the trustworthiness of information that is transferred from IoT objects to the back-end systems. Also, it is important to guarantee that just authorised devices or objects are allowed to do any communication with APIs.
- **Unique Identity:** Giving every device a unique identifier is vital to understanding what the device is, the means by which it acts, different devices it connects with and the proper security measures that should be taken for that device.
- **Encryption:** Strong encryption is critical in ensuring the secure communication between devices. Encryption algorithms should be used to secure data.
- **Internet Security:** It is significantly important to provide security to an IoT network. This can be done by making sure that systems are updated and patched

regularly, providing port security, making sure not to open ports unnecessarily, and using features like firewalls and by detecting and blocking unauthorized IP addresses.

- **Gateways:** Gateways act as an intermediary between the Internet and IoT objects and usually have higher handling force, memory and capacities as compared to the IoT objects themselves. So, it is critical to execute things, like, firewalls that guarantee attackers can't get to the IoT objects.
- **Patch:** It is very important to continuously update software and devices either automatically or over the network connection.

Though IIoT is a mere variant of IoT, IIoT comes with its own security concerns, threats, and vulnerabilities, in addition to inheriting all security concerns of IoT. Besides, if an IIoT incorporates Cloud Computing, it further adds all Cloud-based threats and vulnerabilities. We have discussed Cloud and IoT security concerns above. Further, the additional security concerns specific to IIoT is discussed later.

5.4 IIoT: Security Issues and Challenges

IIoT is a variant of IoT—this implies that IIoT comes with all security concerns incorporated in IoT. In addition to the IoT security discussed earlier, we discuss the security concerns of IoT within industrial context (i.e. IIoT). Security in IIoT is a crucial aspect [18].

Industrial settings may be associated with either product, service or process where IoT can be deployed. The specific security concerns and challenges for IIoT certainly context-specific—the type of industrial application, type of data generated and used within an industrial context, the gadgets used, and the operating and environmental regulations are some of the aspects that influence the security aspects of a given IIoT scenario. Thus, while it is not feasible (if possible, at all) to list all context-specific security concerns by considering millions of different industrial scenarios; it is important to realise the major factors or facets that may act as a driver or source of security concerns for IIoT.

The followings are the key factors influencing IIoT security:

1. **Business Goals and Use of Technology:** An IIoT is implemented with the sole purpose of industrial advantage. Unfortunately, security concerns become a barrier to achieve this. IoT is often associated with Cloud-based technology that introduces Cloud-based IoT. Thus, an IIoT can very well be a Cloud-based IIoT. Thus, the use of technology within an IIoT context needs to be carefully planned and designed to ensure business strategy is complemented by the implementation of IIoT. It is important to keep in mind that IIoT must be design and implemented in line with business goal and strategy. Just the technological advantage and flexibility of IIoT cannot be a justifying reason to implement IIoT, if adverse effect on business is a subsequent possibility. The major challenge to implement IIoT in industrial context from business strategic perspective is that, there is no

"one solution fits all" type of approach—it is a case-by-case basis matter of analysis and decision making on IIoT and its related security trade-offs.

2. **Human Factors:** Human factor plays a very important role in designing and implementing any kinds of computing settings and architectures; IIoT is no exception as well. The related human factors (both generic and those specific to IIoT) are factors that may affect the design and implementation decision of IIoT—this is since human factors create potential security loopholes.

   Human factor related breached are purely dynamic and thus a list of possible breaches resulting from human factors will ever remain non-exhaustive. It can be assumed that the art and science of exploiting human competence (i.e. human factors) to find its new ways to launch security breaches will ever remain as a concern.

   In IIoT, the implementation of digital gadgets needs to be in-line with the level of computer literacy and cybersecurity literacy of the people engaged in an IIoT context. Identification of skill-gap of people involved in IIoT context and proper planning of training and education to minimise the skill-gap (and thus, probability of human factor-centric security breaches) is of utmost importance.

   We argue that the human factor related security concerns and its related analysis must be the topmost priority in design and implementation of IIoT. We perceive human factor related threats to be the most unpredictable and dynamic ones with no exhaustive patterns. Besides, innovative effort in social engineering to exploit human weakness may go unnoticed for long time. Prioritisation. Planning, training and monitoring are required on an ongoing basis. The human factors related concerns applied to all kinds of computing approaches equally, including IIoT.

3. **Human Safety:** From a security perspective, human safety is not to be confused or overlapped with human factors. While human factors refer to the human-centric actions resulting in security breaches or concerns, we stress that human safety (from the perspective of digital security) as the situation arising from using digital technologies (e.g. IIoT) that pose threat to human safety. This is a factor where no tolerance and no degree of 'adjustment' should exist.

   IIoT is promising in industrial context [9], and can involve from information processing to process control. For information processing, IIoT may incorporate all security concerns and threats found in Cloud Computing or other computing means. But when it comes to using IIoT connected machines or devices that controls industrial processes, security concerns and threats related to human safety enter into picture.

   Any IoT enabled industrial apparatus in an IIoT context are theoretically prone to hacking in its traditional meaning. Malware or any other type of attacks may turn an IIoT environment into not only malfunctioning, but also devastating. IIoT comes with the risk of turning industrial non-safety critical systems into malfunctioning safety-critical systems. It would make no credible sense to unnecessarily turn into a non-safety critical system into a malfunctioning safety-critical system—IIoT inherently introduces this risk.

4. **Regulation and Compliance:** Regulation and compliance may emerge as a retrospective security concern due to the design and implementation factors for an IIoT. The relationship among IIoT design strategy and local regulations and compliance are purely context and geographic location based. However, strategic decision impacts security more than technological factors. Thus, the design and implementation of aspects of IIoT demands careful scrutiny. It is crucial for a business to find out all other security concerns, and ask the question, "how regulations and compliance may be violated in case of a breach related to the specific IIoT context?" It is merely impossible to answer this question in general. Every industrial context, in line with business strategy, as well as applicable rules and regulations needs careful justification in this regard. To sum up, the breach of regulations and compliance is not a direct threat, it is rather a secondary risk (in most cases) that results due to other breaches taking place.
5. **Data Storage Location:** IoT is being integrated with Cloud Computing, introducing Cloud-based IoT. Since IIoT is merely a new blend of IoT, it can be well assumed that Cloud-based IIoT also exist and thus the provision and security concerns related to Cloud-based IIoT need consideration. Cloud Computing comes with numerous security concerns. A Cloud-based IIoT thus inherits all security concerns from Cloud Computing in addition to its own threats, vulnerabilities and other security concerns. One biggest issue with Cloud Computing that might emerge as detrimental for Cloud-based IIoT is that, data acquired from an IIoT may be store in remote Cloud data centres not necessarily in the same geographic region. If data is stored in different countries other than the one an industry is located, and in case of any data leakage or breach, the consequence may be in the form of corporate sabotage, breach of privacy of the customers or other stakeholders of the concerned industry, breach of regulations, or in any other form or violation and breach. Cloud Computing may seem to be extremely beneficial from a lot of aspects, but the benefits are simply meaningless and useless if the use of Cloud Computing (and any approach that are based on or incorporated with Cloud Computing) threatens the integrity of an organisation overall robustness and compliance. We have discussed about Cloud Computing security earlier. We argue that the use of Cloud Computing is more of a strategic decision for an organisation than a technological decision; as the adaptation to Cloud Computing is not just the arithmetic sum of the considering Cloud threats as such, but also the indirect strategic impact that may bring to an industry due to using Cloud Computing and any Cloud-incorporated technology (e.g. Cloud-based IIoT).

   If an IIoT context incorporates Cloud Computing, a strategic decision making is required before adopting to the Cloud-based IIoT context. A strategic decision of such migration to Cloud-based IIoT is far more than considering just the technological aspects and security concerns of Cloud Computing—it includes other factors like organisational feasibility and human factors, among other major aspects to consider. An example of factors to be considered to decide on Cloud-based service can be found in [3]. To sum up, Cloud-based IIoT scenario comes with additional security concerns that must be dealt with prior to adopt to such IIoT approach.

6. **Industrial Sabotage:** We mentioned earlier that corporate sabotage is a possibility in IIoT if incorporated with Cloud Computing. To extend further, and to acknowledge its importance, we further discuss this as a topic on its own. The concern regarding industrial sabotage may emerge not only for using Cloud-based IIoT, but also for other factors. IoT, by its definition relies on Internet.

   We recommend the design strategy of a Private IIoT (PIIoT). We define PIIoT as an IoT setting that connects an organisational context without connecting to the Internet. While it can be argued whether exclusion of Internet can still hold an architecture as IoT, it can be safely assumed that an IoT setting with the exclusion of the Internet is much safer than the one connected to the Internet—if a PIIoT can achieve its goal without being connected to the Internet. However, it may not be practical for an IIoT setting to exclude the Internet, but workaround must be in place instead of whimsically connecting an IIoT architecture to the Internet.

   Figure 5.4 illustrates a typical PIIoT scenario. The IIoT is self-contained within the organisation and all the layers of an IoT infrastructure (i.e. Application Layer, Network Layer, Perception Layer) stays within organisational boundary for a PIIoT. Connection and communication to the outside world is through organisational firewall. The communication to and from the outside world should be minimized to the highest possible extent, and batch processing over real-time processing should have priority in information transactions between a PIIoT and the outside world.

7. **Real-time Process Control:** Industrial gadgets and computing enabled equipment may be connected to the Internet through IIoT. Any computing-enabled equipment that are connected to Internet being a part of IIoT opens the possibility for that equipment to be compromised (i.e. hacked) exactly like the way a traditional computer connected to Internet can be compromised. If an IIoT-enabled equipment is compromised, the result may be detrimental. An apparently harmless equipment may become deadly if compromised—hacking is done with no good intention and a hacker can control an industrial equipment not only to remotely operate it in a way to make it malfunction, but also to make it deadly

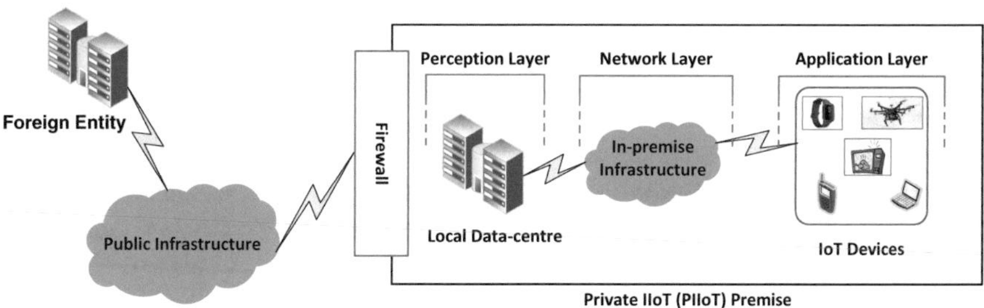

Fig. 5.4 Private IIoT (PIIoT)

to human life. This becomes more crucial for equipment that are ruining real-time processes. Depending on the type of operation, a compromised IIoT-enabled equipment may raise severe safety alarm. Besides, a compromised IIoT-enabled equipment becomes the door for an attacker to collect crucial industrial process-related operational information that may lead to theft of business secrets or intellectual property, resulting in industrial sabotage discussed earlier.

The list of the issues and challenges discussed above are non-exhaustive. when it comes to security aspects of any computing settings or architecture, it is virtually impossible to create a definitive list of issues and challenges from which threats may emerge. Another inherent challenge is to prioritise the discussed factors in terms of severity. For example, we may argue that Industrial Sabotage will have severe impact compared to non-integrity in Regulation and Compliance; however, an industrial sabotage might take place as a retrospective result of non-integrity in regulation and compliance. All the challenges and issues are inter-related, and a loophole in one may adversely affect the integrity of another to subsequently result in breach of security. Besides, the severity of the security issues and challenges are fully context dependent—one factor may be deemed as most severe in one computing context, whereas a different one might be of highest priority for another. One unfortunate aspects of any kind of computing security breach is that, it is not like a dent on a car that become visible (including the reason behind it) right the moment it happens. A security breach may reside hidden for a period of time and the reason behind a detected breach may remain undetected for even longer.

5.5 Discussion and Open Research Problems

The design and implementation challenges for IIoT—when considered from a security perspective—are not only influenced by the IIoT or its context-specific factors. Different Computing architectures often merge or overlap, for example, Cloud-based IIoT incorporates both IoT and Cloud Computing technologies. Subsequently, for a Cloud-based IIoT architecture, security concerns for both IoT and Cloud Computing are to be considered. All different computing aspects are candidate for investigation when it comes to analyse security concerns.

For IIoT, the major design issues need to be perceived both from technological aspects and human factors. Privacy is probably the biggest concerns for Cloud Computing, IoT and IIoT. In addition to these, the potential for a compromised IIoT infrastructure to pose threat to human life becomes another challenge. The computational power of the industrial devices and the feasibility to connect real-time industrial apparatus and processes to any wider network through IIoT are factors that require further investigation. A model or framework for organisations to define safer road-map to adapt IIoT infrastructure is in demand.

Perhaps the major security concern and challenge for organisations to deploy IIoT is not its technical barriers and security concerns arising from technologies.

Rather, the evolution of big data and the increasing extent for organisations to connect to a bigger network (e.g. Internet) through computing approaches (e.g. Cloud Computing, IoT, IIoT) may result in violation of organisational compliance and overall integrity. While apparent security concerns arising from human factors or technological aspects are appreciable, the long term strategic impact on an organisation to adapt to IIoT is a hidden danger, and requires focus and further investigation.

5.6 Concluding Remarks

In this chapter we explored the security issues and challenges of IIoT. While IIoT is derived from IoT, but IIoT has specific goal which aims to provide real-time applications in solving industrial problems. The security concerns for IIoT is thus not only those inherited from IIoT, but also from the business context and the sensitivity of data, process and operation of a business. Having said that, the strategic approach to deploy an IIoT in line with business mission and vision is more important when considering the use of IoT in an industrial context. It is important to realise the technological aspects of an IoT deployment that are associated with security concerns. But more importantly, the strategic decision making and planning in relation to IIoT deployment may bring unknown adverse consequences into an industrial context.

One of the major security concerns in this aspect is to carefully consider all the architectural deployment factors discussed earlier, to avoid or at least to minimise the likelihood of remote malicious parties taking control over industrial operational machines, gadgets, or process. This also applies to preventing eavesdropping and other factors that may result in industrial sabotage. If the implementation of IIoT inevitable, the factors concerning security reasons must be carefully explored, and it lies with the respective industry to decide whether deploying an IIoT is feasible from security perspective.

No compromise should be made with security concerns related to IIoT and the trade-offs needs to be justified from an industry's own context and perspective, instead of accepting IIoT just because of it being a hype, or blindly overestimating only its advantages.

References

1. Ahmed, M.: Ki-Ngā-Kōpuku: a decentralised, distributed security model for cloud computing. Ph.D. Thesis, Auckland University of Technology (2018)
2. Ahmed, M., Litchfield, A.T.: Taxonomy for identification of security issues in cloud computing environments. J. Comput. Inf. Syst. **58**(1), 79–88 (2018)
3. Ahmed, M., Singh, N.: A framework for strategic cloud migration. In: Proceedings of the 2019 5th International Conference on Computing and Artificial Intelligence. ACM, New York (2019), pp. 160–163
4. Alqassem, I., Svetinovic, D.: A taxonomy of security and privacy requirements for the internet of things (IoT). In: 2014 IEEE International Conference on Industrial Engineering and Engineering Management. IEEE, Piscataway (2014), pp. 1244–1248
5. Arregui, D.: Mitigating byod information security risks. In: Australasian Conference on Information Systems (2016)
6. IEEE Standards Association: Internet of Things (IoT) Ecosystem Study. Institute of Electrical and Electronics Engineers, New York (2014)
7. Assunção, M.D., Calheiros, R.N., Bianchi, S., Netto, M.A., Buyya, R.: Big data computing and clouds: trends and future directions. J. Parallel Distrib. Comput. **79**, 3–15 (2015)
8. Azza, A., Hanaa, F., Elmageed, M.A.: IoT perception layer security and privacy. Int. J. Comput. Appl. **975**, 8887 (2019)
9. Bahga, A., Madisetti, V.K.: Blockchain platform for industrial Internet of Things. J. Softw. Eng. Appl. **9**(10), 533 (2016)
10. Bertino, E.: Data security and privacy in the IoT. In: Proceedings of the 19th International Conference on Extending Database Technology (EDBT), vol. 2016 (2016), pp. 1–3
11. Burhan, M., Rehman, R.A., Khan, B., Kim, B.S.: IoT elements, layered architectures and security issues: a comprehensive survey. Sensors **18**(9), 2796 (2018)
12. Buyya, R., Yeo, C.S., Venugopal, S., Broberg, J., Brandic, I.: Cloud computing and emerging it platforms: vision, hype, and reality for delivering computing as the 5th utility. Futur. Gener. Comput. Syst. **25**(6), 599–616 (2009)
13. Che, J., Duan, Y., Zhang, T., Fan, J.: Study on the security models and strategies of cloud computing. Proc. Eng. **23**, 586–593 (2011)
14. Debbabi, B.: Cube: a decentralised architecture-based framework for software self-management. Ph.D. Thesis (2014)
15. Endres, H., Indulska, M., Ghosh, A., Baiyere, A., Broser, S.: Industrial internet of things (IIoT) business model classification. In: Fourtieth International Conference on Information Systems (ICIS) (2019)
16. Garg, A., Rathi, R.: A survey on cloud computing risks and remedies. Int. J. Comput. Appl. **178**(29), 35–37 (2019)
17. Gonzalez, N., Miers, C., Redigolo, F., Simplicio, M., Carvalho, T., Näslund, M., Pourzandi, M.: A quantitative analysis of current security concerns and solutions for cloud computing. J. Cloud Comput. Adv. Syst. Appl. **1**(1), 11 (2012)
18. Haase, B., Labrique, B.: Aucpace: Efficient verifier-based pake protocol tailored for the IIoT. In: IACR Transactions on Cryptographic Hardware and Embedded Systems (2019), pp. 1–48
19. Liu, Y., Tong, K.F., Qiu, X., Liu, Y., Ding, X.: Wireless mesh networks in IoT networks. In: 2017 International Workshop on Electromagnetics: Applications and Student Innovation Competition. IEEE, Piscataway (2017), pp. 183–185
20. Liu, W., Popovski, P., Li, Y., Vucetic, B.: Wireless networked control systems with coding-free data transmission for industrial IoT. (2019, preprint). arXiv:1907.13297
21. Madhu, B., Vaishnavi, K., Dushyanth, N.G., Tushar Jain, S.C.: IoT based home automation system over cloud. Int. J. Trend Sci. Res. Dev. **3**(4), 966–968 (2019)
22. Mell, P., Grance, T., et al.: The NIST definition of cloud computing, SP 800-145 (2011)
23. Perwej, Y., Haq, K., Parwej, F., Mumdouh, M., Hassan, M.: The Internet of Things (IoT) and its application domains. Int. J. Comput. Appl. **182**(49), 36–49 (2019)

24. Razzaq, M.A., Gill, S.H., Qureshi, M.A., Ullah, S.: Security issues in the Internet of Things (IoT): a comprehensive study. Int. J. Adv. Comput. Sci. Appl. **8**(6), 383–388 (2017)
25. Santee, C.D.: An exploratory study of the approach to bring your own device (BYOD) in assuring information security, PhD Thesis, Nova Southeastern University, (2017). Available at: https://nsuworks.nova.edu/gscis_etd/1005/
26. International Telecommunication Union: Series Y: global information infrastructure, internet protocol aspects and next-generation networks next generation networks–frameworks and functional architecture models. International Telecommunication Union: Geneva (2012)
27. Sheikh, A., Munro, M., Budgen, D.: Systematic literature review (SLR) of resource scheduling and security in cloud computing. Int. J. Adv. Comput. Sci. Appl. **10**(4), 35–44 (2019)
28. Venceslau, A., Andrade, R., Vidal, V., Nogueira, T., Pequeno, V.: IoT semantic interoperability: a systematic mapping study. In: International Conference on Enterprise Information Systems, vol. 1. SciTePress, Setúbal (2019), pp. 535–544
29. Wang, L., Von Laszewski, G., Younge, A., He, X., Kunze, M., Tao, J., Fu, C.: Cloud computing: a perspective study. N. Gener. Comput. **28**(2), 137–146 (2010)
30. Wong, K.S., Kim, M.H.: Privacy protection for data-driven smart manufacturing systems. Int. J. Web Services Res. **14**(3), 17–32 (2017)

Chapter 6
Intrusion Detection in Industrial Networks via Data Streaming

Ismail Butun, Magnus Almgren, Vincenzo Gulisano, and Marina Papatriantafilou

6.1 Introduction

The digitalization of society is on-going and affects everything from consumers and their home devices/networks to complex industrial systems such as the *smart grid*. The digitalization brings many advantages, but also comes with risks: due to distributed Internet-of-Things (IoT) architectures, the threat surface is rapidly increasing and many societal critical systems have become susceptible to cyber attacks and even attacked. Anecdotally, whereas cyber attacks and defenses of the twentieth century focused on data and information, attacks of the twenty-first century also focus on physical manipulation of systems with devastating effects [16].

Even though cyber security for traditional IT systems is quite mature, there is a gap when it comes to deploying security mechanisms for industrial networks. The architecture is highly distributed, data is generated at many points in the system and the aggregated volume at central nodes is very big even if data is highly filtered and aggregated beforehand (which may in turn remove important indications of attacks). Orthogonal, there are also aspects of confidentiality and privacy of data which argue for some analysis to be local to the node where data is generated while a full system attack analysis needs to be done centrally.

The explosion of sensor data in the last 20 years has also shown the limitations of databases, where information first needs to be stored and then queried. Driven by general need for analysis of very large data, the *streaming paradigm* was developed emphasizing the *query/analysis* of the data over data persistence in the form of databases. Considering data as flows between nodes makes the data streaming

I. Butun (✉) · M. Almgren · V. Gulisano · M. Papatriantafilou
Network and Systems Division, Department of Computer Science and Engineering, Chalmers University of Technology, Göteborg, Sweden
e-mail: ismail.butun@chalmers.se; magnus.almgren@chalmers.se; vincenzo.gulisano@chalmers.se; ptrianta@chalmers.se

I. Butun (ed.), *Industrial IoT*, https://doi.org/10.1007/978-3-030-42500-5_6

paradigm well suited to the analysis of data in highly distributed industrial networks, such as the smart grid. Data is analyzed on the node and then streamed throughout the system. However, even with the advantages of data streaming for general analysis of data, this same paradigm has not been widely adopted by the security community to enhance security mechanisms tailored to industrial networks.

Following the earlier chapters' presentations of wireless communication needs and challenges, automation trends and applications of Industrial IoT (IIoT) and industrial networks, this chapter is dedicated to present *intrusion detection* and how it can be combined with the principles from *data streaming* to build more efficient attack detection systems for industrial networks.[1]

As data is generated at high rates and volumes in the industrial networks, bottlenecks occur from data/command source to destination, which eventually increases the processing and response times of the queries/commands. In this aspect, we project that the *data streaming paradigm would be a remedy* in helping the efforts of migration of cloud to the edge, meaning that the data are first analyzed as close to the source as possible, before a subset / refinement is sent elsewhere for further analysis. Apart from reducing the volume of data to be analyzed centrally, the local analysis is also faster, meaning that if early signs of attacks are detected the node can quickly recon ure itself to mitigate the attack.

The rest of the chapter is structured as follows: Sect. 6.2 presents the preliminaries by introducing the distributed network architecture of evolving industrial networks along with our use-case scenario, the smart grid. We look at requirements suggested in literature for such networks, as well as give a summary of actual attacks to learn from when designing new detection algorithms. We then introduce *data streaming* in Sect. 6.3 with a short motivating attack detection example. Section 6.4 introduces IDSs, with their underlying detection algorithms with some concrete examples. In Sect. 6.5, we discuss how streaming-based applications can be leveraged to improve intrusion detection systems. Finally, Sect. 6.6 concludes the chapter.

6.2 Preliminaries

This section gives an introduction to the distributed network architecture of evolving industrial networks, followed by an example of such infrastructures, namely a generic smart grid system. The security requirements in such networks are discussed as well as important historical attacks found in the literature.

---

[1] Industrial networks and IIoT are used interchangeably, with a preference for the former term.

6.2.1 Distributed Edge-Fog-Cloud Architectures

Nowadays, with the adoption of edge-computing, the computing resources are approaching to the end-devices from the cloud, hence the term "distributed network" is widely used. We foresee that future industrial network deployments will benefit from this distributed networks wisely. An example to evolving multi-tier architecture of IIoT, Fig. 6.1 [40] presents an abstraction, consisting of:

- Cloud/high-end layer, where processing devices are commonly high-end servers, multi/many -core systems, or supercomputers, such as a *data center* of a business network, in which final data allocations, processing, storage, etc. are handled. Location wise, they are at the farthest point of the network from the low-end layer, yet reachable to/from each end-device.
- Fog/intermediate layer, where processing devices provide moderate computing power, such as edge optimized network servers, located in the vicinity of the low-end layer, in order to provide real-time experiences to the low-end devices.
- Edge/IoT/low-end layer, where resource-constrained on-premise devices are representative examples, such as machines, robot arms, door locks, sensors, etc.

The fog layer constitutes an intermediate layer in between IoT and cloud layers, to provide ways of leveraging cloud layer resources by the resource-constrained IoT end-devices.

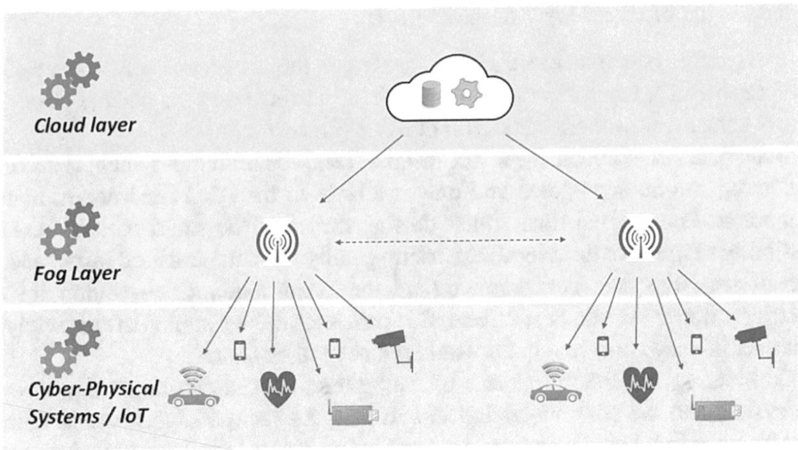

Fig. 6.1 Fog computing and cloud computing layers in IoT networks [40]

6.2.2 A Use-Case Scenario of an Industrial Network: A Smart Grid System

As a use-case scenario of an industrial network, we present a smart grid system. A full grid contains the generation, transmission and the distribution, where we emphasize the electricity distribution network with the coupled information network on top. This part of the grid contains the infrastructure for distributing electricity to customers, from the transmission grid out to the substations and finally the customers with smart meters to measure quality of the delivered electricity as well as to handle billing. The smart meters can be seen as a type of IoT devices, which communicate using ZigBee, BLE, LoRa, NB-IoT, etc. All together, they build up the so called Advanced Metering Infrastructure (AMI), which is the heart of a smart grid system. Typically, we can distinguish the following layers of a smart grid system:

1. **Smart meters:** Instant usage data from the smart meters are collected by the sensors and then sent to the utility head-ends via data concentrators, for further analysis and/or storage to be used at billing, statistics, etc. Think of *IoT layer* of the Fig. 6.1.
2. **Data concentrator:** Data from many (hundreds-thousands) smart meters are collected at the data concentrators and relayed to the AMI head-end. Think of *Fog layer* of the Fig. 6.1.
3. **AMI head-end:** A broad spectrum of the data is stored and processed at the AMI head-end (the utility side), such as the consumer short/long term electricity usage data. Think of *Cloud layer* of the Fig. 6.1.

As a specific example; electricity usage data of the consumers follow the path of 1-2-3 (upstream); whereas commands (such as pushing software patches) from the control center follow the reverse path of 3-2-1 (downstream).

Some data generated at the smart meter is used for billing. As such, it needs to be correct, can be aggregated and only available to the AMI head-end at regular intervals and not in real time. Other data generated at the smart meter offers the possibility to improve the overall electricity quality if it can be shared instantaneous with at least the data concentrator (and the corresponding sub station for the electricity flow). For the best effects, this data should be sampled often (implying customer privacy concerns) but it would not need to be stored.

Considering the attack surfaces of the system, attacks can target any level of the system and the communication in-between. As such, analysis should include data from the individual smart meters (analyze number of local failed logins), some aggregate at intermediate nodes from smart meters to understand if some behavior is wide-spread (a wide-spread attack to many nodes to probe for weak passwords), with the final aggregate analysis in the head-end.

6.2.3 Requirements for Protecting Industrial Networks from cyber-attacks

Mo et al. list the following basic requirements for protecting the users of industrial networks and also the network itself from cyber-attacks [30].

For the Network

- **Integrity of data and commands:** The integrity of the data is critical, as it might affect the operation functions in the factories, the AMI meter readings, the sent commands to the actuators, etc.
- **Availability against DoS/DDoS attacks:** Denial-Of-Service (DoS) attacks cause depletion of the network resources (e.g. throughput, bandwidth, etc.) by sending fake requests either to the server or to the whole network. On the other hand, Distributed DoS (DDoS) attacks are executed by leveraging distributed captured components, such as compromised smart meters of the AMIs; captured firewalls, routers and switches of regular networks; and/or consumer appliances and using them against a single target. DDoS is one of the most dreadful attacks against industrial networks and it is hard to circumvent [47]. For instance, availability of the pricing information and also the power are the key aspects of smart grid networks, therefore they need to be somehow guaranteed (pricing for the provider, and power for the consumer).

For the Users

- **Confidentiality of the usage:** The data including the usage of the services offered by the industrial network provider should be kept confidential. For instance, in a smart grid network, short/long -term electricity usage information of a commercial customer, i.e. an industrial company, should be kept confidential, in order to protect company's production secrets from industrial espionage.
- **Data privacy of the users**: Consumers' private information, such as Personal Identification Information (PII) should not be revealed to the outsiders; an adversary shall not gain any knowledge about individual users of the industrial network without the will of the consumers. Among PII, most sensitive ones are citizen ID number, passport number, passwords of online banking accounts, credit card information, or other financial data, etc. [13]. For the case of smart grid networks, the information pertaining the instant electricity usage patterns might be sensitive as they may reveal personal activities such as presence at the home, being awake or not, etc. All these kinds of sensitive user information should be protected by the network operator (for instance, electricity utility provider in smart grid systems) from unauthorized viewers. More importantly, future implementations of the smart grid systems should comply with the newly released General Data Protection Regulation (GDPR, privacy law of E.U.), by informing the consumers about their data collection processes and obtaining their consent; along with applying transparent data storage/processing policies [36]. Therefore, the utility center should be aware of the total consumption information for billing

operations, yet should decide storing the details of the daily consumption pattern of the individuals according to their privacy choice [3].

Privacy and Confidentiality Goals
As mentioned above, confidentiality of the industrial customers, and privacy of the individual consumers are very important for the industrial networks. Especially after the GDPR law, privacy carries prime importance and if violated, it might be costly for the network operators (such as the utility providers in smart grid systems). While building the industrial networks, particularly following enlisted privacy and confidentiality goals should be aimed at by the network operators [28]:

- *Anonymity*: A user should not be identifiable within a set of subjects.
- *Unlinkability*: After the billing service, any consumption data should not be linkable to the related customer.
- *Undetectability*: The consumption data should not be detectable by adversaries.
- *Unobservability*: An outsider should not observe whether the communication takes place or not regarding execution of certain system related messages and/or other actions of interest, such as sending consumption messages, demand-bidding messages, etc.
- *Pseudonymity*: In smart grid communication, many parties may want to have access to the consumption data from the smart meters or AMI, therefore, a smart meter should have a pseudonym identifier. These pseudonym identifiers can only be possessed by the dedicated entities that are communicating or exchanging messages with the smart meter.

Having these requirements in mind, let's now look at actual attacks that are found in the literature.

6.2.4 Known Cyber-Security Attacks on Industrial Networks

The frequency of cyber-attacks involving industrial networks, especially CPSs, has been expanding. Here, we present only a small sample of well known incidents in chronological order [41]:

Maroochy Shire Sewage Spill (2000) The city council of Maroochy Shire (Queensland, Australia) has outsourced the water treatment facility automation job to a contractor in 1997 and the contractor installed Supervisory Control and Data Acquisition (SCADA—industrial automation and control standard devised by Siemens Inc.) automation tools to the 142 sewage pumping stations. The number of faults recorded has never exceeded two or three per day till late January 2000. However, the number of faults increased drastically whenever cyber-intrusions happened and continued till the date they were identified on 23 April 2000. The pumping stations were normally controlled by the main SCADA station through the dispersed Remote Terminal Unit (RTU)'s. The sabotage was executed by an old employee who took advantage of the wireless RTU's to execute the manipulating commands, an execution of man-in-the-middle (MITM) attack. This overall incident caused a spill of 264K gallons of raw sewage to the environment, which caused an overall bill of $676,000. This is the first ever reported cyber-security incident in the history of SCADA systems [35].

Slammer (2003) Slammer malware targeted a nuclear power plant located at Ohio in 2003 and shut off its safety monitoring system for 5 h. After the incident, it was revealed that the attacker followed a T-1 communication line to connect the corporate network which bypassed the plant's firewall.

Aurora (2007) This vulnerability has shown in 2007 to affect systems that control rotating machinery in the industrial sites such as turbines and diesel generators.

BlackEnergy (2009) This malware targets the Human-Machine Interface (HMI) software of the industrial control systems. It is believed that the first ever hacker-caused power-outage at Ukraine in 2015 was caused by this malware. BlackEnergy was used to steal a legitimate user's Virtual Private Network (VPN) credentials, and by using that attackers gained remote access to the SCADA network of the power distribution and also the HMI. Then they executed further physical attacks such as shutting down the circuits and so on. In a very similar incident at USA, attackers inserted rogue code in software to control electrical turbines of a power plant in 2009.

Stuxnet (2010) This incident is very famous due to political reasons. In 2010, the Stuxnet worm was able to take over many of the Programmable Logic Controllers (PLCs) controlling the centrifuges of the Iranian nuclear facilities, disrupted their centrifuge speed and eventually destroyed them. It has been shown that this worm can be tailored as a platform for attacking smart grid systems that are composed of SCADA systems.

Vampire Attack (2011) Cyber-attacks are sometimes intended to extract secret information, but otherwise, to destroy the communication abilities of the network under attack. For instance, *Vampire Attacks*, in the category of DoS attacks [45], is a very good example of this, in which an attacker intends to drain batteries of the target wireless nodes. In the long run, this type of attack causes a DoS in the overall network; by depleting the batteries of the sensors in the network and causing partition and segregation in the network. These kind of attacks are hard to detect and tough to cope with, and especially dangerous for the IIoT [14].

Havex (2014) It is a malware that uses Remote Access Trojan (RAT) to infiltrate and modify the default software in ICS and SCADA systems. In 2014, it has targeted a number of European companies that develop industrial applications and appliances.

Stealthy Attack (2015) Industrial networks are becoming increasingly susceptible to sophisticated and targeted cyber-attacks initiated by attackers with motivation, domain knowledge, and resources. Recently, a specific kind of attack called *Stealthy Attack* has been discovered to be seriously threatening the industrial environments due to the nature of the attack [29]. The adversaries hide their attacks even at the process level, by injecting just enough malicious data that the compromised sensor values still remain approximately within the noise level. Such stealthy integrity attacks are very tough to detect by anomaly detectors that are not sensitive to noise level fluctuations. Solutions to this kind of attack might require specification-agnostic techniques that monitor time series of sensor measurements for structural changes in their behavior [8]. Coping with this kind of attacks is not easy since they require rigorous tests on carefully crafted attacks in a simulation setting by using the prerecorded data-sets from the selected test-beds with a duration of several days.

Mirai (2016) and Torii (2018) Botnet Attacks Because of lacking rigid security precautions and bad user habits, IoT devices are leveraged as a workforce of the botnets by ill-mannered hackers. For instance, *Mirai malware* is released against Linux OS based IoT devices and aims at gaining shell access of the devices to divert their operations towards the benefit of the *Mirai botnet* [7]. This kind of captured devices are used by the Mirai botnet afterwards in performing joint DDoS attacks toward more advanced targets [18]. *Torii botnet*, is a more sophisticated and an advanced version of Mirai, which also needs to be paid attention while securing industrial networks [32].

6.2.5 *How Can We Enforce Security and Privacy for Industrial Networks?*

Given the requirements discussed in Sect. 6.2.3 and the lessons learned from the attacks listed in Sect. 6.2.4 we now list important general steps to increase the cyber defense.

1. The very first step in increasing the cyber defenses of the industrial networks is executing an extensive security risk analysis of the existing infrastructure, including the research of software, hardware, and communication processes. Furthermore, as intrusions themselves can also provide valuable information, it would be beneficial to analyze system logs and other records of their nature and timing. Already known common weaknesses include poor code quality, improper authentication, and weak firewall rules.
2. Once the first step is completed, then it is suggested to complete an analysis of the potential consequences of the aforementioned failures or shortcomings. This includes both immediate consequences as well as second- and third-order cascading impacts on parallel systems.
3. Thirdly, risk mitigation solutions, which may include simple remediation of infrastructure inadequacies or novel strategies, can be deployed to address the situation. Some such measures include re-coding of control system algorithms to make them more capable of resisting and recovering from cyber-attacks or preventative techniques that allow more efficient detection of unusual or

unauthorized changes to data. Strategies to account for human error which can compromise systems include educating those who work in the field to be wary of suspicious USB drives as in the case of Stuxnet, which can introduce malware if inserted, even if just to check their contents [38].

4. Finally, as a complementary element to the whole security architecture, an **IDS** should be employed, so that in any case of intrusion, system managers can be aware of the threats and take early action against them.

Especially the last point is of interest to us. Even though research of intrusion detection has been ongoing for over 40 years, the current design of such systems may not be optimal for industrial networks. As such, we envision that data streaming (discussed next) can be leveraged for more effective designs (Sect. 6.5).

6.3 Introduction to Data Streaming

We overview in this section the data streaming paradigm. We first discuss its differences from traditional database-based (DB-based) analysis approaches. Subsequently, we introduce basic concepts about the paradigm. Finally, we list some of the available streaming-based analysis frameworks that could be directly leveraged by security applications for industrial systems.

6.3.1 From Database Management Systems to Stream Processing Engines

Database Management Systems (DBMSs) have been used for decades to persist and query data. The amount of data collected in large distributed systems, which has always increased since the introduction of the first pioneer DBMSs, motivated research focusing on distributed and parallel storage and processing. The popularity of DBMSs is not only due to their efficient managing of data, but also due to their powerful language (e.g., SQL), that allows for complex processing of data by means of a set of well known basic commands.

Since the year 2000, nonetheless, the emergence of IoT setups and sensor networks, together with the explosion of the amounts of sensed data, started showing the limitations of DBMSs. Such limitations stem from the fact that the primary goal of DBMSs is data persistence rather than data querying. That is, they are designed to efficiently maintain data that is accessed and aggregated only when a query is issued by a user (or accessed by queries triggered based on user-defined conditions). This data processing paradigm incurs high overheads when applied to applications that are mainly designed to transform and aggregate raw data (possibly coming from unbounded streams) into small, manageable sets of data.

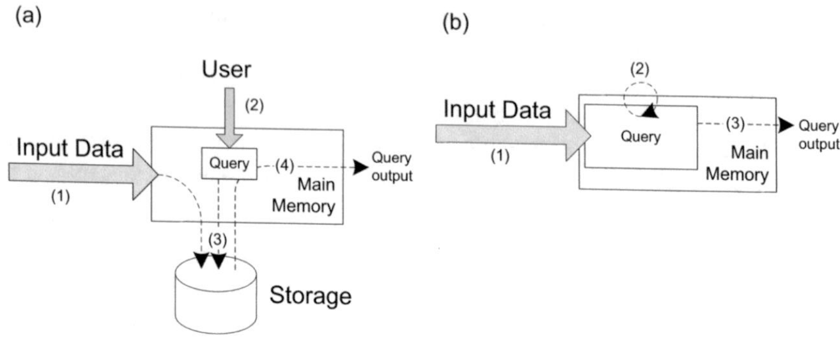

Fig. 6.2 Information processing overview for Database Management Systems (DBMSs) — **a** — and Stream Processing Engines (SPEs) — **b** [22]

To better understand the limitations of DBMSs, Fig. 6.2a (from [22]) presents a high-level overview of how data is stored and processed by a DBMS. Input data is initially persisted into *relations* (1). When a request to process a query from a user is received (2), the query is instantiated and data is read from the storage (3), processed and delivered as output (4). This approach, sometimes referred to as *first the data, then the query* incurs the unnecessary overhead of storing incoming data for applications that are interested in the results of the analysis but that do not require the input data to be persisted once such results are produced.

When the data streaming paradigm was first introduced [39], the main idea was to change the architecture of traditional DBMSs by removing the persistence of each incoming message. The removal of the persistence reduces the per-tuple processing latency significantly as writes to and reads from persistent storage take significantly longer than accesses to main memory.

However, such a modification introduces several new challenges about how data is processed, such as the fact that the available memory is smaller than the available storage space; hence, portions of raw data can only be maintained during limited periods of time. Figure 6.2b (from [22]) overviews how a Stream Processing Engine (SPE) processes data. Input information is processed directly by a continuous query (1). For each incoming message, the query updates its internal state (2). Finally, if it is the case, an output is generated by the continuous query (3).

6.3.2 *Basic Concepts and Sample Application*

A streaming *continuous query* (or simply query, in the remainder) consists of streams and operators. A stream is an unbounded sequence of tuples sharing a schema $\langle ts, a_1, \ldots, a_n \rangle$ where ts is the *timestamp* of the tuple and carries the notion of time for the queries later processing such tuples, and $a_1, \ldots, a_n$ are application-related *attributes*. In a query, tuples are processed by a Directed Acyclic Graph

(DAG) of *operators*, which can also produce new tuples and deliver results to data analysts.

The common operators provided by SPEs includes *Aggregate*, *Join*, *Stateless* and *Merge* operators [22, 23]. Aggregate operators apply aggregation functions over *sliding windows* of tuples. Windows are defined by their size, their advance and, optionally, by a group-by parameter referring to one or more of the input tuples' attributes when the aggregation function is applied independently to each group of tuples sharing such attributes. The Join operator matches tuples from two streams (keeping a sliding window for each stream) and forwards the pairs for which a given predicate holds. Stateless operators, as the name suggests, do not maintain a state evolving with the tuples being processed, and can produce zero, one or more output tuples for each input tuple, applying a user-defined function that specifies the input tuples' attributes to be copied to the output tuples and the functions to be applied to them. Finally, Merge operators allow to merge multiple streams into a single one.

Figure 6.3 shows a sample query intended to monitor the number of login attempts at the terminals of an industrial setup and generates an alert if such number is suspiciously high. More concretely, the application should generate an alert if more than 10 login attempts are made by a user over a period of 5 min. It should be noticed that the suspicious number of attempts can be generated from one as well as from many terminals, and the query should generate outputs that allow to differentiate between the two cases.

The query is composed by Aggregate and Filter operators. In the example, each login monitor is a source of data and generates a tuple composed by attributes ts, id and user (the schema of each stream is shown in grey above each stream). Each tuple specifies which terminal *id* has been used to attempt a login from a given *user* at a given time *ts*. Aggregate A_1 counts the number of distinct login attempts on a per-id, per-user basis, and forwards such count both to Filter F_1 and to Aggregate A_2. Filter F_1 forwards suspicious tuples to the analyst. At the same time, aggregate

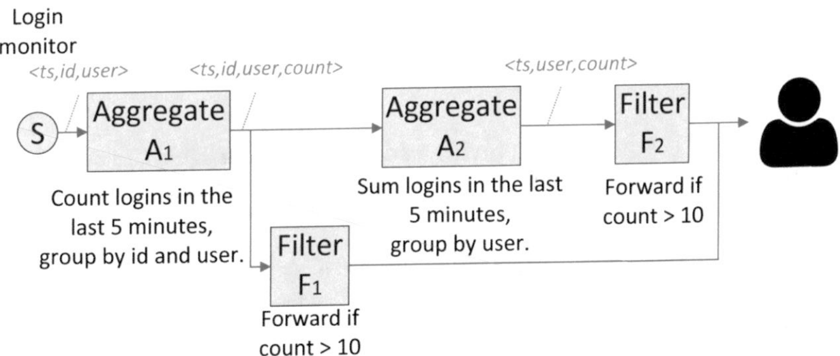

Fig. 6.3 Sample query monitoring the number of login attempts over a distributed set of terminals and generating an alert if more than 10 attempts are made for the same user (either from one or multiple terminals) over a period of 5 min

A_2 produces the cumulative count on a per-user basis. Finally, Aggregate A_2's output tuples are forwarded to filter F_2 which, based on the count, discards them or forwards them to the analyst. The latter can distinguish whether a suspicious alert has been generated from a single terminal or many of them depending on the output stream delivering such alert.

6.3.3 Commonly Used Stream Processing Engines

SPEs have rapidly evolved from research prototypes such as Aurora [1] and Borealis [2] to solutions leveraged by many tech companies. Among existing ones, the most widely adopted include Apache Flink [17], Apache Storm [10], Apache Kafka [27], Apache Beam [9] and Twitter's Heron [26]. While they differ in their architectural choices and the specific APIs they make available to programmers, they usually provide the aforementioned common set of operators for programmers to compose streaming-based applications.

6.4 The Role of Intrusion Detection Systems (IDSs)

Before discussing the special needs of intrusion detection for industrial networks, we will give an overview of the state-of-the art systems used for normal IT systems. As such, Sect. 6.4.1 introduces layered cyber-security life cycle of information systems. IDS categories are introduced in Sect. 6.4.2 and especially anomaly-based IDSs. Specific traditional IDSs that have been used in industrial networks are then summarized in Sect. 6.4.3.

6.4.1 Layered Cyber-Security Life Cycle of Information Systems

As discussed in [12], in order to provide a complete solution against cyber-attacks, any cyber-security system should have a layered defense structure as shown in Fig. 6.4, consisting of prevention, detection and mitigation layers.

1. **Prevention:** This layer, when employed as a system, is referred to as an Intrusion Prevention System (IPS) and constitutes the first line of defense against intrusions. Sometimes, IPSs such as firewalls are not fully trusted and/or they are not efficient enough to prevent all types of attacks towards our networks.

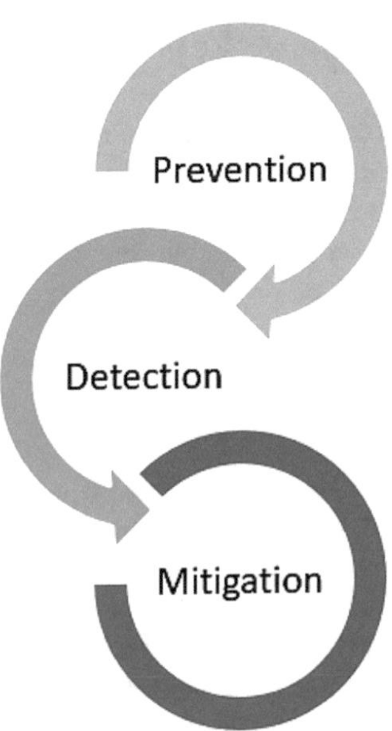

Fig. 6.4 Layered cyber-security life cycle of information systems [14]

2. **Detection:** This layer, when employed as a system, is also referred to as Intrusion Detection System (IDS) and constitutes the second line of defense against intrusions. History has taught us that the first line of defense, IPSs, may fail as in the case of many severe incidents of cyber-attacks against critical infrastructures such as nuclear enrichment facilities, electric grid, etc., as discussed earlier in Sect. 6.2.4. IDSs are complementary for system administrators by offering further solutions to the problem by detecting intrusions to their network on time, so that the threats can be mitigated. Therefore, IDSs are as important as IPSs.
3. **Mitigation:** The last line of defense in the cyber-security is the mitigation, which includes security measures (such as disabling some ports, limiting the Internet access, etc.) to be taken after an intrusion incident is detected.

Even though the methodologies of Prevention/Detection/Mitigation is well-founded in the literature, the terms of IPS and IDS sometimes shift. As an example, we can take Snort (an open-source, free and lightweight network IDS [37]), the de-facto standard when it comes to signature-based IDSs. It started out as an IDS to detect attacks, but later developments allowed it to also react (mitigate) attacks. As such, it is now classified as an intrusion detection and prevention system.

Looking at more specific systems such as industrial networks, vulnerabilities, cyber attacks and the need for IDSs are stressed by Butun et al. [14]. Accordingly, IDS is a very important part of any cyber-security system, and in many cases bears the weight of the overall cyber-defense.

6.4.2 *Intrusion Detection Systems (IDSs)*

As stated earlier in Sect. 6.4.1, Intrusion Detection Systems (IDSs) will constitute the second line of defense against intrusions towards industrial networks. Therefore, it is important to learn their types along with their working principles. In computer science, IDSs can be classified into three categories according to their detection methodology [12, 15, 19]:

1. Anomaly-based IDS,
2. Misuse (signature)-based IDS,
3. Specification-based IDS.

The advantages and disadvantages of IDS types are as shown in Table 6.1.

Commonly, misuse-based systems manually encode indicators of attacks, so-called signatures. As such, these systems are quite specific when they alert in that they can give the type of attack the system is exposed to. These systems work well with very well and categorized attacks of the past. However, they are quite useless in the case of new attack vectors that can not be specified with the old ones.

Specification-based systems are built on the specifications of the allowed behavior of the system and many times used for network protocols. Such systems can be successful when a formal description of the system behavior exists, and that this document is strictly followed. They have been suggested for network protocols, but for very complex systems, a formal specification often does not exist or the behavior of the system is quite dynamic. Specification-based IDS is reported as more suitable to detect process-based attacks, and on the contrary, is observed to be typically expensive for large deployments (such as factory environments) to set up and comparably less scalable [21].

Anomaly-based system are often built using machine-learning techniques in that they try to model the normal system behavior (regardless of its formal specifications). When the current behavior is different *enough* from the learned profile, an alert is generated. Opposed to misuse-based systems, anomaly-based systems are much less exact as they only alert for system anomalies and not true attacks. On the other hand, anomaly-based systems may be able to alert for future

Table 6.1 Comparison of IDS types

IDS type	Advantages	Disadvantages
Anomaly detection-based	Can handle unknown attacks	Low accuracy
	Does not need frequent updates	High false positive ratio
	Easy to configure/generalize	High false negative ratio
Misuse detection-based	High accuracy	Can not handle unknown attacks
	Low false positive/negative ratios	Need frequent updates
Specification detection-based	High accuracy	Hard to design
	Cost-efficient	Hard to generalize
	Low false positive/negative ratios	

unknown attacks (zero-day attacks) if they cause a visible effect in the monitored data [14].

As discussed in [42], it is claimed that the best approach of IDS for industrial networks is *Anomaly-based IDS*, due to its capability of detecting unknown and new types of attacks. Moreover, many industrial networks are quite regular in their behavior (M2M communication), meaning that some of their weaknesses are less pronounced when deployed in industrial networks.

A taxonomy of anomaly detection-based IDSs is shown in Fig. 6.5, and consists of mainly the following approaches: statistical, data mining and Artificial Intelligence (AI) [14]. Each of these approaches is then sub-classified into various methods. The interested reader is referred to [15] for a detailed discussion of these types.

6.4.3 Examples of Traditional IDS Deployed for Industrial Networks

This section presents two recent works focusing on IDS for industrial networks. These approaches employ anomaly-based techniques in an effort of detecting intrusions toward industrial networks but they do not utilize the streaming paradigm to efficiently process data.

6.4.3.1 Example #1

A recent study by Anton et al. [6] on industrial networks has used two different solutions for anomaly detection to capture intrusions:

1. Artificial Intelligence/SVM-Based Solution This solution is based on the *Support Vector Machine (SVM)* which is a machine learning-based anomaly detection method. The *SVM* is a supervised classification and regression analysis method in machine learning which is equipped with learning algorithms to analyze any separated set of data used for classification, meaning the training set needs to contain enough information so that the examples of the different categories are divided by a clear gap that is separating the two groups as wide as possible. New samples are then mapped into the same domain and categorized based on the proximity to the which group they fall near by. It is a large margin classifier and it is trained with a labeled set of instances. The training session is followed by another session, in which the identification and classification of the test and productive data are performed.

2. Data Mining/Data Clustering-Based Solution This solution is based on the *Random Forest* which is a data mining/data clustering and outlier detection-based anomaly detection method. *Random Forest* is a collection of *Decision Trees* which have many binary classifiers comprised of internal split leaf nodes used to classify

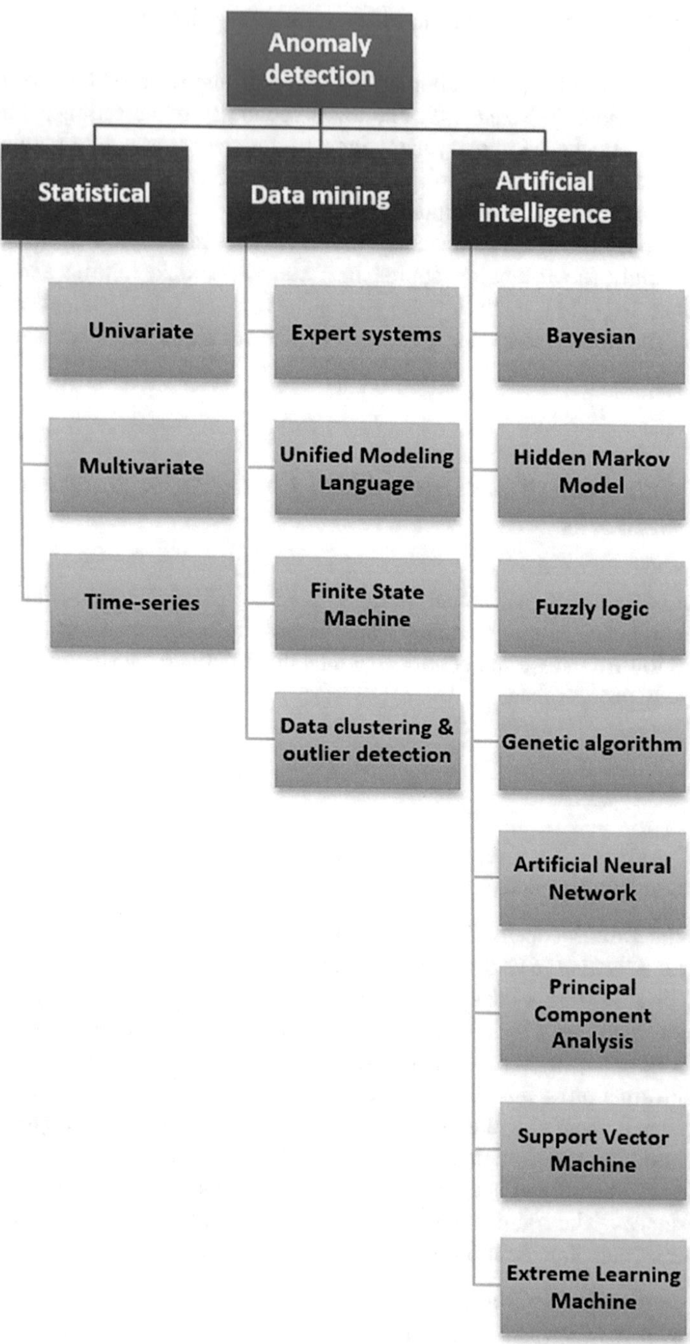

Fig. 6.5 Taxonomy of anomaly detection-based IDSs [14]

Table 6.2 Parameters of the used data sets in [6]

ID	Protocol	# of packets	Duration	Attacks	# of mal. pac.	% of mal. pac.
DS1	Modbus	274,627	4 days	none	60,048	22
DS2	OPC UA	4,910	41 min	2	702	14

mal. pac.: malicious packets

events and bundled together to a one root node. The *majority voting* of all decision trees is the main classification of the Random Forest, hence they are robust to overshooting of the data and can converge to the intended best fit quickly, making them applicable in a variety of use case scenarios [6].

The data containing industrial network in operation is analysed for the sake of discovering the attacks targeted the data. As shown in Table 6.2, two different datasets are employed: Modbus-based gas pipeline control traffic and Object Linking and Embedding for Process Control Unified Architecture (OPC UA)-based batch processing traffic data (from the tanks, pumps, actuators and sensors consisting of water level, flow volume, pressure, temperature, and pump status).

3. Comparison According to the authors, both anomaly detection methods performed well, with the Random Forest method slightly outperforming the SVM method. Methods of machine learning (such as SVM) and data mining (such as Random Forests) can enhance the detection rate of the commercial IDSs in the market. These approaches benefit from a limited state set of the systems as well as a large amount of training data for a given environment, as industrial environments are fruitful data generators.

6.4.3.2 Example #2

In [8], Aoudi et al. introduce the notion of departure-based attack detection where a departure is a specific type of anomaly that refers to the process dynamics being forced to depart from the normal behavior due to potentially malicious structural changes in the stream of sensor measurements. The normal behavior is established in an offline training phase through a mathematical construction that enables the representation of the process dynamics in a noise-reduced geometric space. Thereafter, to detect a departure in the process behavior, devised method computes a departure score during an online detection phase in an iterative way. Whenever this departure score crosses a predetermined threshold, an alarm is raised to the operators.

Aoudi et al. argued that their devised specification-agnostic technique can successfully defend industrial networks against by detecting *stealthy attacks* (see Sect. 6.2.4). This is achieved by monitoring time-series of sensor measurements in the industrial network for structural changes in their behavior.

6.4.3.3 Summary

These two recent examples from the literature show very promising results in detecting attacks in industrial networks. The first example with two methods are using network traffic as input data to find the attacks. As such, a specific wiretap needs to be installed somewhere in Fig. 6.1. If several devices communcicate over the same network link, a single instance may monitor several devices. Otherwise many such taps need to be installed throughout the network. The PASAD algorithm by Aoudi et al. (Example 2) uses sensor data from the process as the input to be analyzed for deviations. As such, it is favorably installed locally at each device. However, the resulting alerts should then be propagated through the system so larger-scale attacks can be detected.

As we discuss in the next section, data streaming for intrusion detection will naturally allow a flow of information throughout the system, making the deployment of Example 1 easier, and allow systems such as described in Example 2 to easily communicate between nodes.

6.5 IDS and Data Streaming

We present in this section why (and how) streaming-based applications can be leveraged for efficient intrusion detection in industrial setups. We begin discussing the different deployment options for the analysis of data in industrial setups. Then, we show how data streaming can enable such deployments by decoupling the semantics of security applications from their distributed and parallel execution. To exemplify this, we also extend the example introduced in Sect. 6.3. Finally, we present some of the streaming-based IDSs discussed in the literature.

6.5.1 What Deployment Options Exist for the Data Analysis?

We can identify three possible hierarchy levels at which data can be processed within an industrial network. In the context of a smart grid, for instance, the options may include following (refer to the details presented in Sect. 6.2.2):

- At the smart meters,
- at the data concentrator,
- at the AMI head-end.

It should be noted that these options are not mutually exclusive. That is, a security application could run both at the meters as well as at the concentrators, as also discussed in [24].

Table 6.3 Comparison of AMI entities [20]

Property	Smart meter	Data concentrator	AMI head-end
Amount of data	Small volume, as data sources are customer's HAN and its associated devices	Large volume, comparatively as it has to handle data from about a few hundred to tens of thousands of smart meters	Huge volume (Big Data), as it has to process data from about several millions of smart meters
Resources[a]	In KB range, are very restrictive	In MB range, more powerful	In GB/TB range, plentiful resources composed of high-end servers
Data speed	Comparatively low, because of non-frequent requests at the smart meter	High, as it aggregates a good number of smart meters' data	Very high, as it needs to handle a huge amount of meter data, event data, commands, etc.

[a]Main memory, processor capacity, etc

Table 6.3 is inspired from [20] and provides a good comparison of all entities in the smart grid systems, especially from the data streaming point of view. As we travel from the smart meter to the AMI head-end (leaf to the root), the amount of data to be handled increases drastically, as well as the speed required to carry and process that data. This is a clear indication of where the data streaming can be used and why.

6.5.2 *Leveraging Distributed and Parallel Execution of Streaming Applications in Industrial Setups*

As we discussed in Sect. 6.3, a streaming application is a DAG of base operators composed into a continuous query. Once the semantics of a certain streaming application have been expressed by composing such base operators, parallel execution is achieved by (1) instantiating multiple copies of each operator and enabling data-parallelism by distributing their input data (according to the semantics of the operator) and then collecting the results, as well as by (2) assigning such operators to distinct threads (possibly belonging to different processes and/or nodes).

Figure 6.6 shows a possible deployment plan for the sample query introduced in Fig. 6.3. In this example, we assume an industrial setup is composed of a collection of remote terminals and a set of servers that, together, compose a data center used for the analysis of gathered data. The idea is to run the query presented in Fig. 6.3, which monitors the number of login attempts over a distributed set of terminals and generates an alert if more than 10 attempts are made for the same user (either from one or multiple terminals) over a period of 5 min, in a parallel and distributed fashion, making the overall analysis scalable by distributing as much as possible data at the edge rather than gathering all of it in the central data center. In the sample deployment, a pair of operators A_1 and F_1 are deployed at each remote terminal to locally check if more than 10 login attempts are made by each user. If that is the case, the output produced by operator F_1 (which is merged by a merger operator M at a different node that, in this example, is a server at a central data center) is then forwarded to the end user. At the same time, the partial counts forwarded by all instances of operator A_1 are also merged at the data center server and the cumulative sums, if exceeding the threshold 10, are also forwarded as alerts.

Leveraging of the data streaming paradigm and SPEs allows for monitoring applications to be deployed at all the levels of a given industrial setup hierarchy by means of distributed and parallel operator execution [31, 34]. Distributed execution (achieved by means of *inter-operator parallelism*) allows for operators belonging to the same query to be run at different nodes (e.g., A_1 and A_2 in Fig. 6.6). At the same time, parallel execution (achieved by means of *intra-operator parallelism*) allows for individual operators to be run in parallel at arbitrary numbers of nodes (e.g., A_1 and F_1 in Fig. 6.6). We refer the reader to [22] for a exhaustive discussion about the parallelization of data streaming operators.

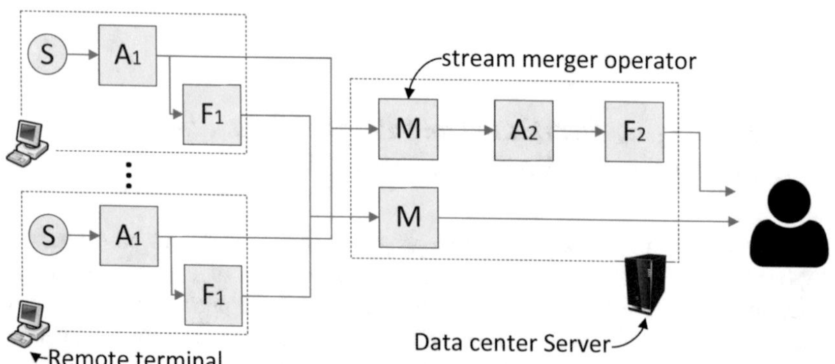

Fig. 6.6 Distributed and parallel deployment of the sample query from Fig. 6.3. The query monitors the number of login attempts over a distributed set of terminals and generates an alert if more than 10 attempts are made for the same user (either from one or multiple terminals) over a period of 5 min

6.5.3 Correctness Guarantees Enabled by the Data Streaming Processing Paradigm

The execution of a query is said to be *deterministic* if feeding the query the same sequence of input tuples results in the query producing the same sequence of output tuples. For monitoring applications in CPSs, determinism is crucial to achieve correct and predictable/repeatable analysis, especially when the latter is used to generate sensitive or safety-related alerts [33].

A way to achieve such behavior is to compose queries with base operators that, if fed the same sequence of input tuples, produce the same sequence of output tuples [22, 46]. That is, operators that have no randomness in their analysis and whose analysis does not depend on execution-specific aspects [4, 22]. The common way for stateful operators to achieve deterministic execution is to base the latter on the notion of time carried by the tuples' timestamps (also referred to as *event* time [4, 17]).[2] Continuing the example of Fig. 6.6, this implies that, once each remote terminal generates a tuple about a user login and sets a timestamp to it, an alert will be generated (if the threshold condition is satisfied) independently of the latency and interleaving of messages at the data center server.

6.5.4 Applications of the Streaming Paradigm in the Context of IDSs in the Literature

Even though streaming has been developed since 2000 and offers a structured way to consider parallel and distributed analysis of data over complex systems such as the smart grid [11, 25, 43, 44], few examples of intrusion detection systems that use the technique exists in the literature. Below we summarize two promising examples to highlight the possibilities.

METIS Gulisano et al. [24] propose METIS. METIS relies on probabilistic models for detection and is designed to detect challenging attacks in which the adversaries aim at going unnoticed. Owing to its two-tier architecture, it eases the modeling of possible adversary goals and allows for a fully distributed and parallel traffic analysis through the data streaming processing paradigm. At the same time, it allows for complementary intrusion detection systems to be integrated in the framework.

Gulisano et al. have shown METIS' use and functionality through an energy exfiltration attack scenario, in which an adversary aims at stealing energy information from AMI users. Based on a prototype implementation using the Apache Storm SPE and a very large dataset from a real-world AMI, authors have shown

[2]In this case, a common assumption in the literature is that all nodes of a distributed setup have synchronized clocks [23].

that METIS is not only able to detect such attacks, but that it can also handle large volumes of data even when run on commodity hardware.

Stream Data Mining for Distributed-IDS Alseiari et al. [5] propose a real-time Distributed Intrusion Detection System (DIDS) for the AMI infrastructure that utilizes stream data mining techniques with a multi-layer implementation approach. Using unsupervised online clustering technique, the anomaly-based DIDS monitors the data flow in the AMI and distinguishes if there is anomalous traffic. The authors employed *Mini-Batch K-means* clustering technique which is a variant of "K-means" clustering algorithm, that is suited to data stream models.

It incrementally fits the new sessions to the existing model by continuously learning and updating the clusters in real time which significantly reduces running and computation time. However, K-means follows a static procedure by discarding the previously created clusters and building a new model from scratch in every *sliding window* iteration, hence using more memory and time. By executing a comparison between online and offline clustering techniques, the authors claim that their experimental results have shown that online clustering "Mini-Batch K-means" algorithm was able to suit the architecture requirements by giving high detection rate and low false positive rates.

6.5.5 Security Implications and Opportunities of Data Streaming

As in every complex system, inclusion of new components (as well as removal of the old ones) might provide opportunities along with implications. As such, inclusion of *data streaming* paradigm will have pros and cons against the already existing components, users, etc, as discussed below.

Security Implications This is a debated issue, whether centralize v.s. distributed detection of intrusions is more critical and beneficial. It is argued in this chapter that distributed and real-time monitoring of events (intrusions) is important for industrial networks, especially for mission-critical systems, such as smart grid systems presented in Sect. 6.2.2, hence timely taken decisions might save people from catastrophic events.

In most instances, bottlenecks (the data traffic gets congested!) occur in the upstream path of the centralized networks. However, distributed networks are resilient to that problem. For instance, intrusion detection solutions (especially related to the data streaming) discussed in this chapter, aim at pushing the data analysis towards peripheral devices instead of gathering data centrally. This kind of solution is offered to provide a fast response by removing the necessity of the round-trip messages in between the smart meter and the server. Besides, data streaming is also beneficial to the overall network performance in distributed networks due to the decreased load in preventing the aforementioned bottlenecks.

On the other hand, centralized solutions have shown to be more effective against more dedicated and distributed attacks such as DoS and DDoS attacks.

Opportunities This chapter argued and also discussed that the *data streaming* paradigm can be really helpful for industrial networks in fulfilling detection of intrusions timely manner, hence it allows continuous monitoring of events in an autonomous and adaptive way. It can be also beneficial in decreasing the bottlenecks in big industrial networks such as smart grid systems presented in Sect. 6.2.2, by not only enhancing the response time of the queries/commands but also decreasing the overall traffic of the network.

6.6 Conclusions

This chapter presented key points of industrial networks, especially network architecture, data handling (centralized vs. distributed), and cyber-security vulnerabilities, attacks and their counter measures. Besides, data stream processing as means to analyze data in IIoT infrastructures is described. Overall research challenges are also presented regarding cyber-security aspects of data streaming for industrial networks, especially for smart grid systems. Finally, applicability of data streaming to IDSs is proven by the evidence from the literature that appeared in the recent years.

Based on the recent developments in the field, it is concluded in this chapter that the *data streaming* concept can be utilized for industrial networks, especially for smart grid systems, by leveraging cloud and/or fog based deployments. More importantly, *data streaming* is projected to be very useful and handy for detecting intrusions in timely manner and cost efficient way.

Acknowledgments This research has been partially supported by the Swedish Civil Contingencies Agency (MSB) through the projects RICS, by the EU Horizon 2020 Framework Programme under grant agreement 773717, and by the Swedish Foundation for International Cooperation in Research and Higher Education (STINT) Initiation Grants program under grant agreement IB2019-8185.

References

1. Abadi, D.J., Carney, D., Çetintemel, U., Cherniack, M., Convey, C., Lee, S., Stonebraker, M., Tatbul, N., Zdonik, S.: Aurora: a new model and architecture for data stream management. VLDB J. **12**(2), 120–139 (2003)
2. Abadi, D.J., Ahmad, Y., Balazinska, M., Cetintemel, U., Cherniack, M., Hwang, J.H., Lindner, W., Maskey, A., Rasin, A., Ryvkina, E., et al.: The design of the borealis stream processing engine. In: Second Biennial Conference on Innovative Data Systems Research (CIDR), vol. 5 (2005), pp. 277–289
3. Abdallah, A., Shen, X.: Security and Privacy in Smart Grid. Springer, Berlin (2018)
4. Akidau, T., Bradshaw, R., Chambers, C., Chernyak, S., Fernández-Moctezuma, R.J., Lax, R., McVeety, S., Mills, D., Perry, F., Schmidt, E., et al.: The dataflow model: a practical approach to balancing correctness, latency, and cost in massive-scale, unbounded, out-of-order data processing. Proc. VLDB Endowment **8**(12), 1792–1803 (2015)
5. Alseiari, F.A.A., Aung, Z.: Real-time anomaly-based distributed intrusion detection systems for advanced metering infrastructure utilizing stream data mining. In: 2015 International Conference on Smart Grid and Clean Energy Technologies (ICSGCE). IEEE, Piscataway (2015), pp. 148–153
6. Anton, S.D.D., Sinha, S., Schotten, H.D.: Anomaly-based intrusion detection in industrial data with SVM and random forests. (2019, preprint). arXiv:1907.10374
7. Antonakakis, M., April, T., Bailey, M., Bernhard, M., Bursztein, E., Cochran, J., Durumeric, Z., Halderman, J.A., Invernizzi, L., Kallitsis, M., et al.: Understanding the mirai botnet. In: USENIX Security Symposium (2017), pp. 1092–1110
8. Aoudi, W., Iturbe, M., Almgren, M.: Truth will out: departure-based process-level detection of stealthy attacks on control systems. In: Proceedings of the 2018 ACM SIGSAC Conference on Computer and Communications Security. ACM, New York (2018), pp. 817–831
9. Apache Beam (2016). https://beam.apache.org/. Accessed 25 Sept 2019
10. Apache Storm (2017). http://storm.apache.org/. Accessed 25 Sept 2019
11. Botev, V., Almgren, M., Gulisano, V., Landsiedel, O., Papatriantafilou, M., van Rooij, J.: Detecting non-technical energy losses through structural periodic patterns in AMI data. In: 2016 IEEE International Conference on Big Data (Big Data), pp. 3121–3130. IEEE, Piscataway (2016)
12. Butun, I.: Prevention and detection of intrusions in wireless sensor networks. University of South Florida, Ph.D. Thesis (2013)
13. Butun, I.: Privacy and trust relations in Internet of Things from the user point of view. In: 2017 IEEE 7th Annual Computing and Communication Workshop and Conference (CCWC). IEEE, Piscataway (2017), pp. 1–5
14. Butun, I., Österberg, P.: Detecting intrusions in cyber-physical systems of smart cities: challenges and directions. In: Secure Cyber-Physical Systems for Smart Cities, pp. 74–102. IGI Global, Pennsylvania (2019)
15. Butun, I., Morgera, S.D., Sankar, R.: A survey of intrusion detection systems in wireless sensor networks. IEEE Commun. Surveys Tutorials **16**(1), 266–282 (2013)
16. Butun, I., Österberg, P., Song, H.: Security of the Internet of Things: vulnerabilities, attacks and countermeasures. IEEE Commun. Surv. Tutorials **22**, 616–644 (2019)
17. Carbone, P., Katsifodimos, A., Ewen, S., Markl, V., Haridi, S., Tzoumas, K.: Apache flink: stream and batch processing in a single engine. Bull. IEEE Comput. Soc. Tech. Committee Data Eng. **36**(4), 28–38 (2015)
18. Chaudhary, S.: Privacy and security issues in Internet of Things. Int. Edu. Res. J. **3**(5), 2433–2436 (2017)
19. Debar, H., Dacier, M., Wespi, A.: Towards a taxonomy of intrusion-detection systems. Comput. Netw. **31**(8), 805–822 (1999)
20. Faisal, M.A., Aung, Z., Williams, J.R., Sanchez, A.: Data-stream-based intrusion detection system for advanced metering infrastructure in smart grid: a feasibility study. IEEE Syst. J. **9**(1), 31–44 (2014)

21. Fauri, D., Dos Santos, D.R., Costante, E., den Hartog, J., Etalle, S., Tonetta, S.: From system specification to anomaly detection (and back). In: Proceedings of the 2017 Workshop on Cyber-Physical Systems Security and PrivaCy. ACM, New York (2017), pp. 13–24
22. Gulisano, V.: Streamcloud: an elastic parallel-distributed stream processing engine. Ph.D. Thesis, Universidad Politécnica de Madrid (2012)
23. Gulisano, V., Jimenez-Peris, R., Patino-Martinez, M., Soriente, C., Valduriez, P.: Streamcloud: an elastic and scalable data streaming system. IEEE Trans. Parallel Distrib. Syst. **23**(12), 2351–2365 (2012)
24. Gulisano, V., Almgren, M., Papatriantafilou, M.: Metis: a two-tier intrusion detection system for advanced metering infrastructures. In: International Conference on Security and Privacy in Communication Networks. Springer, Berlin (2014), pp. 51–68
25. Gulisano, V., Almgren, M., Papatriantafilou, M.: When smart cities meet big data. Smart Cities **1**(98), 40 (2014)
26. HERON: A realtime, distributed, fault-tolerant stream processing engine from Twitter. https://apache.github.io/incubator-heron/. Accessed 30 Sept 2019
27. Kafka Streams. https://kafka.apache.org/documentation/streams/. Accessed 10 Sept 2019
28. Kumar, P., Lin, Y., Bai, G., Paverd, A., Dong, J.S., Martin, A.: Smart grid metering networks: a survey on security, privacy and open research issues. IEEE Commun. Surv. Tutorials **21**, 2886–2927 (2019)
29. Mo, Y., Sinopoli, B.: On the performance degradation of cyber-physical systems under stealthy integrity attacks. IEEE Trans. Autom. Control **61**(9), 2618–2624 (2015)
30. Mo, Y., Kim, T.H.J., Brancik, K., Dickinson, D., Lee, H., Perrig, A., Sinopoli, B.: Cyber–physical security of a smart grid infrastructure. Proc. IEEE **100**(1), 195–209 (2011)
31. Najdataei, H., Nikolakopoulos, Y., Papatriantafilou, M., Tsigas, P., Gulisano, V.: Stretch: scalable and elastic deterministic streaming analysis with virtual shared-nothing parallelism. In: Proceedings of the 13th ACM International Conference on Distributed and Event-Based Systems. ACM, New York (2019), pp. 7–18
32. Osborne, C.: Meet torii, a new IoT botnet far more sophisticated than mirai variants (2018). https://www.zdnet.com/article/meet-torii-a-new-iot-botnet-far-more-sophisticated-than-mirai/. Accessed 10 Sept 2018
33. Palyvos-Giannas, D., Gulisano, V., Papatriantafilou, M.: Genealog: fine-grained data streaming provenance at the edge. In: Proceedings of the 19th International Middleware Conference. ACM, New York (2018), pp. 227–238
34. Palyvos-Giannas, D., Gulisano, V., Papatriantafilou, M.: Haren: a framework for ad-hoc thread scheduling policies for data streaming applications. In: Proceedings of the 13th ACM International Conference on Distributed and Event-Based Systems. ACM, New York (2019), pp. 19–30
35. Sayfayn, N., Madnick, S.: Cybersafety analysis of the maroochy shire sewage spill, working paper cisl# 2017-09. Cybersecurity Interdisciplinary Systems Laboratory (CISL), Sloan School of Management, Massachusetts Institute of Technology (2017), pp. 2017–09
36. Sharma, R.: How does GDPR affect smart grids? (2018). https://www.energycentral.com/c/iu/how-does-gdpr-affect-smart-grids
37. Snort: Snort network intrusion detection system. https://www.snort.org. Accessed 5 Nov 2019
38. Sridhar, S., Hahn, A., Govindarasu, M.: Cyber–physical system security for the electric power grid. Proc. IEEE **100**(1), 210–224 (2011)
39. Stonebraker, M., Çetintemel, U., Zdonik, S.: The 8 requirements of real-time stream processing. ACM Sigmod Rec. **34**(4), 42–47 (2005)
40. Stylianopoulos, C.: Parallel and distributed processing in the context of fog computing: high throughput pattern matching and distributed monitoring. Licentiate Thesis, Chalmers University of Technology (2018)
41. Tesfay, T.T.: Cybersecurity solutions for active power distribution networks. Ph.D. Thesis, EPFL (2017)

42. Tong, W., Lu, L., Li, Z., Lin, J., Jin, X.: A survey on intrusion detection system for advanced metering infrastructure. In: 2016 Sixth International Conference on Instrumentation & Measurement, Computer, Communication and Control (IMCCC). IEEE, Piscataway (2016), pp. 33–37
43. van Rooij, J., Gulisano, V., Papatriantafilou, M.: Locovolt: Distributed detection of broken meters in smart grids through stream processing. In: Proceedings of the 12th ACM International Conference on Distributed and Event-based Systems. ACM, New York (2018), pp. 171–182
44. van Rooij, J., Swetzén, J., Gulisano, V., Almgren, M., Papatriantafilou, M.: eChIDNA: continuous data validation in advanced metering infrastructures. In: 2018 IEEE International Energy Conference (ENERGYCON). IEEE, Piscataway (2018), pp. 1–6
45. Vasserman, E.Y., Hopper, N.: Vampire attacks: draining life from wireless ad hoc sensor networks. IEEE Trans. Mobile Comput. **12**(2), 318–332 (2011)
46. Walulya, I., Palyvos-Giannas, D., Nikolakopoulos, Y., Gulisano, V., Papatriantafilou, M., Tsigas, P.: Viper: a module for communication-layer determinism and scaling in low-latency stream processing. Future Generation Comput. Syst. **88**, 297–308 (2018). https://doi.org/10.1016/j.future.2018.05.067
47. Yan, Q., Huang, W., Luo, X., Gong, Q., Yu, F.R.: A multi-level DDoS mitigation framework for the industrial Internet of Things. IEEE Commun. Mag. **56**(2), 30–36 (2018)

Index

I. Butun (ed.), *Industrial IoT*, https://doi.org/10.1007/978-3-030-42500-5

Printed in the United States
by Baker & Taylor Publisher Services